SECRETS OF SACRED ECOLOGY

How to Create Peaceful, Magical, and Transformative Spaces in Nature

CHUCK PETTIS

2025
Earth Sanctuary
1385 Raden Place
Freeland, WA 98249 USA

First Edition: 2025

Secrets of Sacred Ecology: How to Create Peaceful, Magical, and Transformative Spaces in Nature

Cover photo: Credit Pawel Gaul, iStock

Identifiers:

ISBN 979-8-9925312-0-6 (print)

ISBN 979-8-9925312-1-3 (eBook)

Dedication:
Gratitude to the Constellation

To the bright constellation of all who gather
In the sacred circle of this place, Earth Sanctuary—
The family who hold me with steady hands,
Friends whose smiles dance through the trees,
Lamas, teachers, wisdom of the ages,
Guiding each step like stars aligning in the dark.

To authors whose words have shaped my soul,
Advisors and consultants, trusted minds,
Architects of visions, builders of dreams,
Gardeners sowing seeds of patience and peace,
Designers crafting beauty in every leaf.

To employees, service providers, hands that serve,
CPAs, attorneys, numbers, and laws made kind,
Ecologists and birders who see the Earth's heartbeat,
Neighbors whose kindness ripples like a stream,
Retreatants and visitors, seekers of stillness,
Members whose presence turns solitude into community.

And to the unseen, the spiritual beings
Who walk in silence but speak to the heart—
Gratitude is the thread that binds us all,
A luminous weave of love and care,
Together, we create this living sanctuary,
A place of peace, a place of light.

Thank you, each and every soul,
For your gifts, your time, your spirit shared.
In this constellation, we all shine bright,
Together, a universe of love and life.

Chuck Pettis — visionary, author, and environmental artist—has designed and built sacred spaces according to the tenets of art, science, and spirituality since 1970. Chuck Pettis is the author of the book, *Secrets of Sacred Space: Discover and Create Places of Power*. Chuck Pettis was recognized in the September 2008 issue of Science of Mind magazine as one of 12 people making a difference in the world.

A long-time environmentalist, Chuck decided to take personal responsibility for enhancing biodiversity and creating a sustainable sacred ecology by creating Earth Sanctuary (https://earthsanctuary.org/), a sacred space sculpture garden, nature reserve, and retreat center. With a 500-year plan to restore the 72 acres of Earth Sanctuary to a mature old-growth forest with a maximum diversity of birds, wildlife, plants, and fungi, Earth Sanctuary is an exemplary model for sacred ecology design.

Earth Sanctuary is noted for its sacred spaces designed for meditation and prayer. These sites bring together ancient art forms, a profound connection with nature, and diverse global spiritual traditions.

Chuck Pettis is president of Brand-Solutions (https://brand-solutions.com/) and author of *TechnoBrands: How to Create & Use "Brand Identity" to Market, Advertising & Sell Technology Products.*

Chuck Pettis has a B.A. in Psychology from Carnegie-Mellon University, Pittsburgh, Pennsylvania, and an M.S. in Design from Southern Illinois University, Carbondale, Illinois.

Contents

Preface

In 1975, I visited the Rollright Stone Circle near Oxford, England, and was profoundly moved by the palpable power of this ancient sacred space. Standing among the stones, I felt a deep connection to something timeless and magical. That moment ignited a lifelong quest to understand how to create places imbued with such transformative energy.

In 1999, after decades of research and exploration, I published *Secrets of Sacred Space: Discover and Create Places of Power*. The book shared my insights about the art, science, and spirituality inherent in sacred spaces. But I yearned to do more than write—I wanted to bring these principles to life.

In 2000, I purchased 72 acres of woodland and three ponds and began creating Earth Sanctuary, a sacred space sculpture garden, nature reserve, and retreat center. By blending scientific knowledge, spiritual practices, and artistic expression, I transformed the property into a living experiment in sacred ecology.

From the very beginning, Earth Sanctuary visitors shared stories of profound, magical experiences. Over time, the property's energy deepened, enveloping everything in a sense of peace and connection. Trees and plants thrived, wildlife became more visible, and visitors found themselves captivated by moments of unexpected grace—a silent owl perched in the Cottonwood Stone Circle, or a deer pausing to meet their gaze before disappearing into the forest.

Something miraculous has happened. Earth Sanctuary has become a sacred ecology, a harmonious balance of nature and spirit. This book, *Secrets of Sacred Ecology: How to Create Peaceful, Magical, and Transformative Spaces in Nature*, weaves together actionable guidance, ancient wisdom, and personal stories from my pilgrimage to discover the secrets of creating places of peace, magic, and transformative experiences in nature.

-Chuck Pettis

Introduction to Sacred Ecology

My name is Chuck Pettis, and I am the founder and creator of Earth Sanctuary, a sacred space sculpture garden, nature reserve, and retreat center on Whidbey Island, near Seattle, Washington, USA.

This book is the story of my journey, spiritual practices, and the techniques I've used to create a place where nature takes center stage—a magical space where people experience peace, inspiration, and transformation. Through this book, I'll share how Earth Sanctuary was conceived and built, guiding you to use these same practices and techniques in your own garden, backyard, or community space.

For a quick overview of Earth Sanctuary and the love, care, and spirit infused into this sacred ecology, watch "Discover Earth Sanctuary" (https://tinyurl.com/3nh3ac3p), a video by Whidbey Telecom.

When I purchased the Earth Sanctuary land 25 years ago, I had three goals:

- Create an old-growth forest.
- Create a sacred space sculpture garden based on my book *Secrets of Sacred Space.*
- Create a retreat center.

Accomplishments Over the Past 25 Years

Earth Sanctuary has grown into a sanctuary for both nature and spirit. Here's what we've achieved:

- **Ecological Restoration**: Over 15,000 native plants (80 species) and 3,350 trees (35 species) have been planted, creating habitats for over 100 bird species, including Hooded Mergansers, Wood Ducks, Barred Owls, Great Horned Owls, and Ospreys.

- **Sacred Spaces**: Eleven powerful sacred spaces now grace the land, including a Dolmen, Labyrinth, two Medicine Wheels, three Stone Circles, spiritual sculptures, a Tibetan Buddhist Stupa, and the Infinite Tower Sculpture.
- **Retreat Center**: Our retreat center has hosted over 1,000 spiritual retreats, ranging from one-day to 100-day experiences. The center benefits from the accumulated spiritual practices of past retreatants and continues to bless new visitors. A special thanks to Celia Sullivan, Earth Sanctuary's caretaker, whose dedication ensures a nurturing space for spiritual practice.

Additionally, over 100 Memorial Trees have been planted to honor loved ones, often with ashes interred nearby, creating a living legacy.

But most importantly, Earth Sanctuary has become a place of transformation. Visitors report deeply spiritual, magical, and healing experiences, describing it as:

- An enchanted forest.
- One of the most healing places I have ever been.
- A powerful place with magical energy.
- Earth Sanctuary is truly a sacred place.
- It is very evident that the earth here is holding the loving energy for all that visit.
- This place is the embodiment of inspired intuition!
- A sacred land where the magnificence is palpable.

What is Sacred Ecology?

Something has happened at Earth Sanctuary in the past 25 years that is quite extraordinary. I call it *Sacred Ecology*.

> ***Sacred ecology explores the interconnected relationships between the Six Kingdoms of Life, Spiritual Beings, and their environment, to foster the most diverse, vibrant, and healthy ecosystems imaginable.***

The book, *Secrets of Sacred Ecology: How to Create Peaceful, Magical, and Transformational Spaces in Nature* explains the principles and practices that have made Earth Sanctuary what it is today. It serves as a guide for anyone seeking to heal the Earth and create their sacred ecology, whether large or small. Consider this book a manual for building and sustaining a nurturing, successful, and sustainable relationship with nature.

What to Expect from This Book

The book is divided into two parts:

> **Part 1: How I Learned the Secrets of Sacred Ecology**: Covers the importance of meditation, ley line power centers, ceremonies, and spiritual beings, as well as methods for creating diverse ecosystems that attract birds and wildlife.
>
> **Part Two: Earth Sanctuary's Sacred Spaces**: Chronicles the creation of Earth Sanctuary's sacred spaces, explaining their symbolism, meaning, and power.

In Part One – How I Learned the Secrets of Sacred Ecology, I will explain the powers and importance of meditation, ley line power centers, ceremony, and spiritual beings. I will tell you how to connect and communicate with trees. I will share how to create a diverse and healthy ecosystem that attracts birds and wildlife. In Part Two, I will tell the stories of how I created each of Earth Sanctuary's sacred spaces and their symbolism, meaning, and power.

Each chapter includes questions to help stimulate a rich and meaningful group discussion.

My passion for sacred spaces and sacred ecology started in the mid-1970s when I visited the United Kingdom and felt the power of ancient monuments like the Rollright Stone Circle, Stonehenge, Avebury, Newgrange, Callanish, Maes Howe, and more. These places moved me and enabled me to practice deeper meditations, getting to peace and clarity faster. I resolved to find the techniques used to create places of power. After 20 years of study, my book *Secrets of Sacred Space: Discover and Create Places of Power* (https://a.co/d/9rBbVyp) was published in 1999 by Llewellyn Publications. In 2000, I purchased 72 acres of woodlands, ponds, and wetlands and began creating Earth Sanctuary.

Who This Book Is For

This book is dedicated to anyone who:

- Loves nature and seeks to enhance the natural world.

- Cares deeply about the environment and wants to make a positive impact locally.
- Has visited Earth Sanctuary and wants to learn more about its creation.
- Desires to create their own sacred space, no matter the size.
- Feels the need for a place of safety and spiritual renewal.
- Recognizes global challenges such as climate change and ecological destruction and wishes to take personal action.
- Strives to create an uplifting, healing, and sustainable sacred ecology.

What is Sacred?

Spirituality, at its core, is an individual's pursuit of the sacred and their purpose in life. Sacredness refers to something consecrated, set apart, and made holy through association with spiritual practices, blessings, or ceremonies.

Bridging Ecology and Spirituality

Ecology explores the interconnected relationships among living organisms and their environment. *Sacred Ecology* goes a step further, recognizing the presence of spiritual beings—such as Deities, God, Angels, Buddhas, and Creator—and their role in shaping our world.

By integrating ecological restoration with spirituality, art, and ceremony, sacred ecology fosters peace, magic, and biodiversity, creating sanctuaries for all living beings.

Characteristics and Criteria of a Sacred Ecology

- Peaceful
- Magical
- Spiritual
- See and hear birds and wildlife
- Ongoing ecological restoration to increase biodiversity
- An atmosphere that evokes mindfulness, meditation, and prayer
- Symbolic sacred spaces located in places of power
- Ceremonies and blessings from spiritual and religious leaders and masters
- The feeling of being in a heaven on earth with angels and other spiritual beings

Secrets of Sacred Ecology – Chapter Overview

In this book, I explain what I have done to create not only a sacred space, but a sacred ecology that creates an atmosphere of peace, magic, love, happiness, and relief from the stresses of everyday life that is also a diverse and healthy ecosystem. I hope that you can benefit from my experience and adapt what I do to your situation.

In Chapter One, "The Power of Meditation," I explain the power of meditation to grow your awareness of the metaphysical world around you and empower you to create sacred spaces and connect with spiritual beings that bless all living beings.

In Chapter Two, "The Power of Ley Lines and Ley Line Power Centers," you will learn of the underground Earth Energies that hold and emanate the energies, emotions, and spiritual experiences of people who visit and meditate at sacred spaces. This chapter is further explored in Part Two of the book, which describes the sacred spaces at Earth Sanctuary.

In Chapter Three, "The Power of Spiritual Beings," I introduce you to the Spiritual Hierarchy and how they can help you create ley line power centers, make decisions, and deepen your spiritual practice.

In Chapter Four, "The Power of Ceremony to Bless and Heal" and Chapter Five, "The Power of the Stupa, Prayer Wheels, and Prayer Flags" you will learn the benefits of inviting spiritual and religious leaders and masters to conduct ceremonies and blessings and how prayer flags and prayer wheels can radiate love, happiness, and the relief of suffering.

Creating an atmosphere of safety and sanctuary is vital, as described in Chapter Six, "The Power of Space Itself to Alter Consciousness."

The importance of planting long-living trees and preserving existing old and tall trees as well as the techniques for communicating with trees is explained in Chapter Seven, "Connect and Communicate with Trees," supported by the Earth Sanctuary Memorial Tree Program covered in Chapter Eight.

Think long-term! In Chapter Nine, "The Power of a Diverse and Healthy Ecosystem," I explain how to design with nature and our 500-year plan to create an old-growth forest. I also cover the importance of diversity, recount my adventures with beavers, and how we are attracting Wood Ducks and Hooded Mergansers with nesting boxes.

The Relationship Between Spiritual Practice and Ecology

For caretakers of a sacred ecology, a strong foundation in spiritual practice is invaluable. Regular meditation fosters mindfulness, clarity, and a deeper connection to the metaphysical and natural realms. Practicing loving-kindness and mental silence enhances one's ability to:

- Connect with the spiritual beings present in nature.
- Make thoughtful, intuitive decisions that honor the land.
- Cultivate an atmosphere of peace, renewal, and transformation for all who visit.

Spiritual practice also amplifies the blessings and energy infused into the sacred spaces. As visitors engage with these spaces, they, too, benefit from the positive vibrations and leave with a deeper appreciation for the interconnectedness of all life.

The Solution to Biodiversity Is Sacred Ecology

Sacred ecology is a comprehensive conservation effort focused on restoring sustainable biodiversity and ecosystem health. It combines exemplary restoration ecology with art and spirit to create a sanctuary for all living beings and a peaceful place for personal renewal and spiritual connection.

Since 2000, this has been my vision statement: *Earth Sanctuary combines exemplary ecology with art and spirit to create a sanctuary for birds and wildlife and a peaceful place for personal renewal and spiritual connection.*

Sacred ecology emphasizes spiritual practices, deities, symbolism, subjectivity, reverence, and earth energies while building on scientific ecology, which uses traditional scientific approaches to knowledge using empirical observation, experimentation, and modeling.

In sacred ecology, we recognize and honor the interdependence of all species of the six kingdoms and work to create the optimal environment for each of them with minimal harm and ecological destruction. My goal is for all beings small-to-large to be happy and free from suffering.

The six kingdoms of life represent the major categories into which all living organisms are classified based on shared characteristics. Here is an overview of the six kingdoms:

1. **Archaea**
    - Organisms: Archaea (also called archaebacteria)

- Characteristics: Unicellular, prokaryotic (no nucleus), often thrive in extreme environments (e.g., hot springs, salt lakes, deep-sea vents), and have unique biochemical properties.

2. **Bacteria**
   - Organisms: Bacteria (also called eubacteria)
   - Characteristics: Unicellular, prokaryotic, found in diverse environments, can be beneficial (e.g., gut bacteria) or pathogenic (cause disease).
3. **Protista**
   - Organisms: Protists
   - Characteristics: Mostly unicellular (some multicellular), eukaryotic (have a nucleus), highly diverse group including algae, amoebas, and slime molds.
4. **Fungi**
   - Organisms: Fungi
   - Characteristics: Mostly multicellular (except for yeasts), eukaryotic, obtain nutrients by decomposing organic material (e.g., mushrooms, molds, and yeasts).
5. **Plantae**
   - Organisms: Plants
   - Characteristics: Multicellular, eukaryotic, autotrophic (produce their own food through photosynthesis), including trees, flowers, and grasses.
6. **Animalia**
   - Organisms: Animals
   - Characteristics: Multicellular, eukaryotic, heterotrophic (consume other organisms for food), including humans, mammals, birds, insects, and more.

Sacred Ecology Tactics and Best Practices

- Make Nature the Top Priority
- Focus on a property, garden, backyard, or community space with maximum potential for biodiversity.

- Combine restoration ecology with art and spirit to create a sanctuary for all living beings and a peaceful place for personal renewal and spiritual connection.
- When outside in Nature, practice invisibility by being aware, quiet, and respecting wildlife.
- Design with Nature: Create and implement a long-term plan to create a diverse, healthy, and sustainable ecosystem.
    - Honor and protect existing large trees
    - Plant long-living, drought-resistant trees and plants with mycelium. Mycelium is the "wood wide web" of the forest, incredibly tiny "threads" of a greater fungal organism that connect trees and plants for the greater good of the fungi and the plants and trees.
    - Conduct ongoing ecological restoration to increase biodiversity.
    - Recognize and honor the interdependence of all species of the five kingdoms and try to create the optimal environment for them with minimal harm and ecological destruction.
    - For a large property, refer to The Book of Wilding (https://a.co/d/01XUbRU)
- Sponsor ceremonies and blessings from spiritual masters and religious leaders to bless and heal the property
- Create sacred spaces over ley line power centers using the techniques in *Secrets of Sacred Space*.
- Encourage visitors and friends to meditate, pray, and practice mindfulness while on the property and especially in sacred spaces
- Connect, communicate, associate, and cooperate with the spiritual beings of your spiritual path or religion.
- Act for the happiness and relief of suffering of all beings, from humans to the most minute forms of life.
- Ethical behavior while on the property:
    - Treat the land, other visitors, and wildlife with respect and love.
    - Avoid disturbing the birds and wildlife on the property.
    - Walk quietly and speak softly.
    - Stay on the marked trails.

Some Practical Recommendations For Larger Properties

- Generate income to sustain the sacred ecology.

  - Admission fee: An admission fee is needed to increase the perceived value of the sacred ecology and to provide income to sustain the sacred ecology.
  - Membership Programs: Offer annual memberships to individuals and families. Members can receive benefits such as unlimited access (Annual Pass program), invitations to exclusive events, and newsletters detailing updates and progress.
  - Sponsorship Opportunities: Provide options for donors to sponsor specific elements, such as planting trees, maintaining sacred spaces, or supporting wildlife conservation initiatives. At Earth Sanctuary we have a Memorial tree program.
  - Merchandise and Products: Sell eco-friendly merchandise and educational materials to serve as a lasting reminder for visitors of their connection to the space.
  - Workshops and guided tours.
  - Honor system: With two entrances and being open every day during daylight hours, Earth Sanctuary has to operate on the honor system. We have had some problems with money being stolen from our payment boxes. Our solution is a welded metal box with a narrow slit for cash and envelopes. Since most locks in hardware stores can be easily opened by lockpicking, we use high-end security locks, e.g., Abus Granit 37/55.

- No dogs:
  - Birds and other animals think of dogs as predators and the presence of one dog, even leashed, will disrupt their normal behavior.
  - Deer activity is significantly reduced in trails in areas that allow dogs.
  - Small mammals, including squirrels, rabbits, chipmunks, and mice also exhibit reduced levels of activity in proximity to trails in areas with dogs.
  - Dogs are avid chasers and can displace wildlife from their habitats.
  - Dogs can spread diseases: canine distemper, rabies, parvovirus, plague, giardia, and muscle cysts.
  - Dogs can damage the area set aside for sensitive native plants and animals

- No trash cans[1] or picnic tables. At Earth Sanctuary, when I had a trash can, people filled it up. I took the trash can away and after that, no one left trash. Picnic tables take away from the peace and quiet that invites birds and wildlife. If appropriate, create a separate place for a store, café, restrooms, etc.

Final Thoughts and Call to Action

This book is more than a guide—it is an invitation and a call to action. At its core, *Secrets of Sacred Ecology* is about rekindling humanity's sacred relationship with nature.

Whether you care for a small garden or a vast expanse of wilderness, the principles shared here can empower you to:

- Create spaces that heal, inspire, and transform.
- Contribute to restoring biodiversity and combating ecological destruction.
- Connect with the spiritual essence of the natural world.

Let us work together to honor and protect Mother Earth. May this journey fill you with a sense of purpose, joy, and connection as you take steps to create your sacred ecology. The time to act is now. Let your love for nature guide you—for the benefit of all beings, present and future.

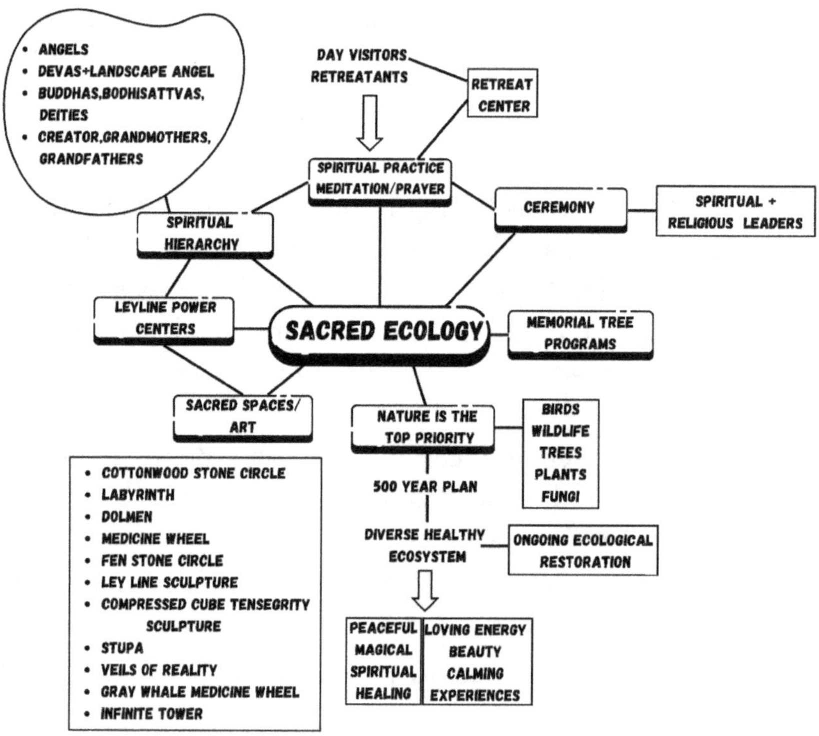

Figure 1- Earth Sanctuary's Sacred Ecology Mind Map (c) Chuck Pettis

Questions for Group Discussions

These questions should help stimulate a rich discussion on the various aspects of Earth Sanctuary and its broader implications on spirituality, ecology, and personal growth.

- What are your initial impressions of Chuck Pettis' vision and accomplishments at Earth Sanctuary?
- Have you ever visited a place that felt sacred or profoundly impacted you?
- How does the concept of "Sacred Ecology" resonate with you? Do you think it can play a role in addressing ecological and environmental issues?
- Which of the testimonials from visitors stood out to you? Why do you think Earth Sanctuary has such a profound effect on people?
- How important do you think it is for places like Earth Sanctuary to exist in today's world?
- What steps could you take to create a small-scale sacred ecology in your community or backyard?

Part 1

How I Learned the Secrets of Sacred Ecology

Chapter 1:
The Power of Meditation and Spiritual Practice

Meditation is the key, the foundation for creating a sacred ecology, and our area of focus for Chapter 1. If you have not already done so, I encourage everyone to find their spiritual path or religion and make meditation and prayer part of your life. For those of you who don't have a meditation practice, I have included a short meditation manual below, followed by my journey of meditation and spiritual practice

A Short Meditation Manual[2]

The purpose of meditation is to develop the ability to:

- Focus the mind.
- Become aware of your body and your thoughts.
- Become aware of your inner feelings and emotions.
- Become aware of the spiritual beings all around us.
- Understand the source and nature of happiness and suffering.
- Understand your mind and how it works.
- Increase your mindfulness to nip impatience and anger in the bud.
- Be peaceful and happy with yourself.

The Seven-Point Meditation Posture

- Sit in a full-lotus position, half-lotus (right foot on the outside, left on the inside), or cross-legged. If sitting on the floor is difficult, sit on a chair.
- Keep the back straight.
- Hands on the lap – right hand on the left hand with palms up and thumbs touching.
- Place the tip of the tongue on the upper roof of the mouth behind the upper teeth, with the mouth slightly open.

- Eyes gazing gently past the nose, unwaveringly.
- The neck is slightly bent forward, chin slightly down, shoulders slightly back, and chest slightly out.
- Do not move.

Length of Meditation Sessions

- Start with short daily meditation sessions. 60-second meditations can be very effective. It is better to have one short meditation practice every day than a long one weekly.
- Observe your progress and the improvement in your ability to calm and focus your mind for longer time periods.

Six Simple Meditation Techniques

1. Observe Breath
2. Count Breaths
3. Count 21 Breaths
4. Observe Thoughts
5. Count Thoughts
6. Mantra Recitation – Om Mani Padme Hung

1. Observe Your Breath

- Sit in meditation posture.
- The object of your concentration is your breath.
- Breathe normally.
- Observe the inhalation and exhalation of the breath.
- Don't pay attention to how you inhale and exhale. Just watch the breath coming in and going out.

2. Count Your Breaths for 60 Seconds

- Sit in meditation posture.
- The object of your concentration is your breath.
- Observe the inhalation and exhalation of the breath.
- Count how many breaths you have. One breath equals one inhale and one exhale.

- I attended the Dalai Lama's 80th birthday celebration in New York City. The Dalai Lama recommended that everyone do this breath-counting meditation.

3. Count to 21 Breaths

- Sit in meditation posture.
- The object of your concentration is your breath.
- Observe the inhalation and exhalation of the breath.
- Count to 21 breaths. One breath equals one inhale and one exhale.
- After you can count to 21 breaths without getting distracted, see how many breaths you can count without getting distracted.

4. Observe Your Thoughts

- Sit in meditation posture.
- The object of your concentration is your thoughts.
- Observe your thoughts.
- Watch the thoughts come and go.

5. Count Your Thoughts

- Sit in meditation posture.
- The object of your concentration is your thoughts.
- Observe your thoughts.
- Count how many thoughts you have in 60 seconds.

6. Radiate Love

- We all have an imaginary "love bucket." We all want our love bucket to be filled with love.
- Bring to mind the feeling of love. To do that, think of someone or something that evokes the feeling of love in your heart.
- Think of someone who needs extra love right now.
- Send your love to them for 60 seconds. Notice that the love in your love bucket doesn't go down. When you do this spiritual practice, you have primed your "love pump" and the love keeps coming!

7. Recite Prayers or Mantras

Various religions and spiritual traditions incorporate the recitation of prayers or mantras as a key spiritual practice. Here are some notable examples:

Native Americans and Indigenous Peoples

- American Indian spirituality is nature-centered, there is no separation between creation and humans. Creation includes people, wildlife, fish, birds, mountains, the wind, water, and everything in the physical and spiritual realms. All creation is prayed for, and all physical reality has spirits in charge of each aspect. The creator and spirits must be honored and acknowledged with our prayers
- Ceremonies led or sanctioned by holy people are the place to pray for help to the spirits, creator, and ancestors. Ceremonies include:
    - Sweat lodge, vision quest, sun dance, and pipe ceremonies.
    - Prayers, dancing, singing, drumming, chanting
    - Medicine Wheel ceremonies with prayer flags, drumming, and chanting

Hinduism

- Mantras: Reciting mantras is central to Hindu practice. Common mantras include the "Gayatri Mantra" and "Om Namah Shivaya."
- Prayers: Various prayers and hymns, like the "Vishnu Sahasranama" and "Hanuman Chalisa," are recited.

Buddhism

- Mantras: Tibetan Buddhism is known for mantras such as "Om Mani Padme Hung," a prayer that all beings may be happy and free from suffering. One can recite the mantra Om Mani Padme Hung at any time --while walking, sitting, driving, or working.
- In many Buddhist traditions, chanting sutras is also common.
- Prayers: Buddhists often recite prayers, such as the "Refuge Prayer" and dedications of merit.

Christianity

- Prayers: Christian practice includes a wide range of prayers, such as the "Lord's Prayer," "Hail Mary," and the "Rosary" in Catholicism.
- Chants: Gregorian chants and other liturgical hymns are recited, especially in Catholic, Orthodox, and Anglican traditions.

Islam

- Prayers: Muslims perform Salah (ritual prayers) five times a day, which include recitations from the Quran.

- Dhikr: Repetition of divine names and phrases, such as "Subhanallah," "Alhamdulillah," and "Allahu Akbar."

Sikhism

- Prayers: Sikhs recite prayers from the Guru Granth Sahib, such as the "Japji Sahib" and "Rehras Sahib."
- Mantras: The "Mool Mantar" and "Waheguru" mantras are commonly recited.

Jainism

- Prayers and Mantras: Jains recite the "Namokar Mantra" and various prayers for purification and spiritual advancement.

Taoism

- Prayers and Chants: Taoists often chant texts from the "Tao Teh Ching" and other Taoist scriptures.

If you seriously want to create a sacred ecology, here are my recommendations:

- Deepen your spiritual practice.
- Train your mind's ability to focus clearly and one-pointedly on an object of meditation without effort. Develop the ability to breathe 10 breaths without a thought[3].
- Learn how to connect with the spiritual beings of your spiritual path or religion.

Overview of this Chapter

In this chapter, I will share my journey into a lifetime of meditation and spiritual practice, which has enabled me to create sacred spaces and connect with spiritual beings that bless all living creatures at Earth Sanctuary. I will recount how I learned to dowse for water wells, water lines, ley lines, and ley line power centers. You will discover the first stone circle I built in 1977—the Ellis Hollow Stone Circle in Ithaca, NY. Dorothy MacLean of Findhorn introduced me to the realm of Devas and spiritual beings, while John and Kate Payne guided me in clearing spaces of negative energies, ghosts, and spirits. His Holiness Jigdal Dagchen Sakya initiated me into the profound spiritual practices of Tibetan Buddhism, which I continue to practice daily.

I have also undertaken two pilgrimages to the sacred Buddhist sites of India and Nepal. A month at Ghoom Monastery near Darjeeling, India, allowed me to receive the precious Lamdre teachings—the path to enlightenment. These experiences and teachings have profoundly blessed me, empowering me to cultivate a spiritual ecology that brings happiness and relief from suffering to the visitors and retreatants of Earth Sanctuary.

The Beginning of My Spiritual Career

Early Encounters with Religion

My parents took my sister and me to the Presbyterian Church as a child. By mutual consent, after a short time, we stopped going. My sister and I did not find it inspiring or relevant to our young lives. My mother decided she would rather sleep late than get up and go to church. I remember hard pews, cards to doodle on, and boredom. My father later became an exemplary Christian, attending a Baptist Church in Macon, Georgia. When I visited him and my mother there, I attended church with them. When my Dad visited Seattle, he sat in the back at Sakya Monastery of Tibetan Buddhism. We supported each other's spiritual path.

The Night That Changed Music Forever

The night that changed music forever: The Beatles' American Debut on The Ed Sullivan Show. On February 9, 1964, my Dad, Mom, and sister were all sitting in front of the TV when the Beatles took to the stage for their first televised performance in America, forever altering the course of music history ... and my life. I played their first US album, Meet the Beatles, over and over. I am listening to that album now. I can sing along with all the words, 60 years later (just like the Taylor Swift fans of today)!

The Beatles inspired me to start playing guitar and growing my hair. In Junior and Senior High, I played in bands: The Citations, Tony and the Tigers, and the Fabulous Rancid Yaks (some indication of my future connection to Tibet). We played at high school dances, parties, and fraternities. What fun we had.

Music as a Spiritual Gateway

Music and sound are profound gateways to spirituality, resonating with the deepest parts of our being. They transcend words, reaching directly into the soul to evoke emotions, awaken memories, and create a sense of connection to something greater than ourselves. From the harmonious vibrations of Tibetan singing bowls to the rhythmic chants of sacred mantras, sound carries the power to heal, uplift, and transform. It aligns our energy, bridges the physical and the ethereal, and reminds us of the universal rhythm that underlies all existence. In this way, music and sound become sacred tools for meditation, celebration, and communion with the divine.

College, War, and New Values

Fast forward to college and the Vietnam War. I was on student deferment at Carnegie-Mellon University in Pittsburgh, but my high school friends were getting drafted. One of them, Richard Ryan, the guitar player in The Citations, was killed in Vietnam. I had some serious thinking to do about my values. I decided that killing people was not something I could do and joined the Quakers, The Society of Friends, is a Christian group that rejects outward rites and an ordained ministry. They have a long tradition of actively working for peace and opposing war. With their help, I became a conscious objector.

On December 1, 1969, a lottery drawing – the first since 1942 – was held to determine who would be sent to the Vietnam War. Men with lower numbers were called first. The lower the number the higher the chance of going to Vietnam. The highest lottery number called for my group was 195; My number was 312. I celebrated like a big weight had been lifted from my shoulders. I was relieved because I knew I wouldn't make a good soldier, and I didn't want to die in a war that didn't make sense and was wrong.

In college, I played in two bands – "The Plant" and "Tiger Rose and the Fabulous Thunderbirds." There is a magic that happens when a band and the audience get in sync. Musicians know that feeling well and want more of it! An energy that goes back and forth and builds. The heat rises. The music, the band, and the dancers become one. For musicians, this is the best high. You go beyond yourself, there is no thinking, no separation.

A Spiritual Awakening at Carnegie-Mellon University

I had some spiritual experiences of sorts at Carnegie-Mellon. One night in front of Skibo, the student union, I looked up at the night sky and felt something bigger than myself. I asked myself. What did the future hold for me? What was my little role in this huge Universe? Did I have a destiny? And then I heard Buckminster Fuller speak and I became inspired to help make the world work for all humanity.

Buckminster Fuller Leads Me to Meditation

The most important person in my school life at Carnegie-Mellon was John Paskiewicz, a design professor. He became my independent study professor and introduced me to Buckminster Fuller, inventor of the geodesic dome, tensegrity (along with Kenneth Snelson), and the World Game: "Make the world work, for 100% of humanity, in the shortest possible time, through spontaneous cooperation, without ecological offense or the disadvantage of anyone." Buckminster Fuller's World Game was an educational simulation intended to create solutions for overpopulation and the uneven distribution of global resources. The World Game was an interactive role-playing experience where participants would organize and distribute world resources, e.g. food, energy, etc. Players worked non-competitively to allocate the earth's natural capital worldwide, scoring points through efficient management. I read all of Fuller's books at the time and heard him speak at least twice – he could talk nonstop for hours!

John encouraged me to go to graduate school at Buckminster Fuller's Design Department at Southern Illinois University. I came to play Buckminster Fuller's World Game at the SIU Design School, seen as the second Bauhaus of design innovation.

Discovering Mediation

I never played the World Game. Instead, everyone I met was meditating. For some reason, Carbondale was a major center for Ananda Marga Yoga, one of the first Indian spiritual groups to come to the US. I was ready for meditation. It was like my karma drew me to meditation and in a big way. When you join a spiritual group, you get an initiation, a ceremonial ritual, that connects you to the teacher and starts you on a spiritual path. I was initiated by Acharya Vimalananda and received a mantra, special words repeated in harmony with your breath to aid concentration in meditation. A bright white light blossomed in my "third eye." That got my attention!

We had weekly one-hour group meditations and I began meditating two hours a day. Looking back on it now, I am amazed I could meditate that long. This was something that my spiritual self was attracted to. It felt natural and peaceful. I was asked, where were the immediate benefits and effects on my life? When you are sitting quietly contemplating your breath, thoughts, feelings, and sensations arise. If you have said something hurtful to another, it comes up and you want to avoid that feeling. So, slowly I became a better person, morally and ethically. I still work on that every day.

Life in Ananda Marga

I was invited to be on the National Board of Ananda Marga. We drove several times to Wichita in my Ford Supervan filled with our friends for big retreats with hundreds of people meditating and doing Kirtan. We sang the Baba Nam Kevalam mantra while ideating on the Supreme Being. Kirtan inculcates the feeling of bliss and prepares the mind for meditation, which is done after the singing.

My Ananda Marga meditation teacher was Acarya Yatiishvarananda Avadhuta (Dadaji). He was a true sanyasi, living the life of a monk, with supreme ethics and spiritual practice. He initiated me into the higher spiritual practices of Ananda Marga. Years later, he would visit me in Seattle and also visit the Earth Sanctuary labyrinth.

According to Ananda Marga tradition, I got a Sanskrit spiritual name. My Ananda Marga name was Dharmapala, "protector of the Dharma." It turns out that Dharmapala is also a Buddhist divinity who, though benevolent, is represented as hideous and ferocious in order to instill terror in evil spirits. Perhaps a precursor to my later work to clear spaces of negative energy and entities/ghosts.

Figure 2- Acarya Yatiishvarananda Avadhuta with my sons, Morgan and Bre circa 1997. (c) Chuck Pettis

Service to Humanity

A key part of Ananda Marga was service to humanity. My wife at the time, Sina, and I became big brothers and sisters of two kids who lived in a ramshackle two-room house with about 10 other people, all sharing the same bedroom. Water came from an outdoor faucet. The social worker told us that in her practice in Southern Illinois, she had encountered much poverty. There is so much suffering in the world.

Dadaji told me he wanted to work with inmates, so we went to the Marion Federal Penitentiary where he initiated dozens of inmates in the chapel. Every man there was in for life or more. I went there every week to lead group meditations. Arriving at the penitentiary was surreal. The grounds were filled with deer. It took seven doors to get into the center where we did meditations. It was an intense place. A powerful place to meditate!

Building a Geodesic Dome

In my first year at Southern Illinois University, I built a 3-frequency 5/8 geodesic dome south of Carbondale. Two friends, Bill and Kathy, lived in this dome. There were no rectangular corners or chairs. Storage, seating, and working spaces were truncated tetrahedrons. The acoustics were very clear – you could hear everything. Every month on the full moon we would have a group meditation in the dome. The Mylar windows faced South. The moon would seem to move through the arcs during the evening.

Bill and Kathy moved to Hawaii where my current wife, Claudia, and I once visited them. They were living in geodesic domes in the forest and swimming with dolphins every day.

Figure 3- Carbondale Dome 1971. (c) Chuck Pettis

Figure 4- Truncated tetrahedrons inside the dome. (c) Chuck Pettis

The Art of Letterpress Printing

In my second year at SIU, I started a private press, Bliss Press, and hand-printed Lao Tzu's *Tao Teh Ching: The Bible of the Loving Road* by Robert Finley. I handmade paper and hand-bound the book in Chinese style. This became my master's thesis: the hand-printed and bound book and a short slide show showing how it was done. This was in 1972, before copy machines. My advisor, "Mr. G" told me to "make a beautiful box" to hold the carousel and book. When I walked into the grad school office to submit my thesis, the woman in charge of accepting theses said, "What a beautiful box." Thesis ACCEPTED! This was one of my first learnings about branding: packaging is important. The book cover matters!

Figure 5- Setting type for the Tao Teh Ching. (c) Chuck Pettis

Figure 6- Making paper for the book cover (c) Chuck Pettis

Figure 7- The Bible of the Loving Road - Tao Teh Ching book cover. (c) Chuck Pettis

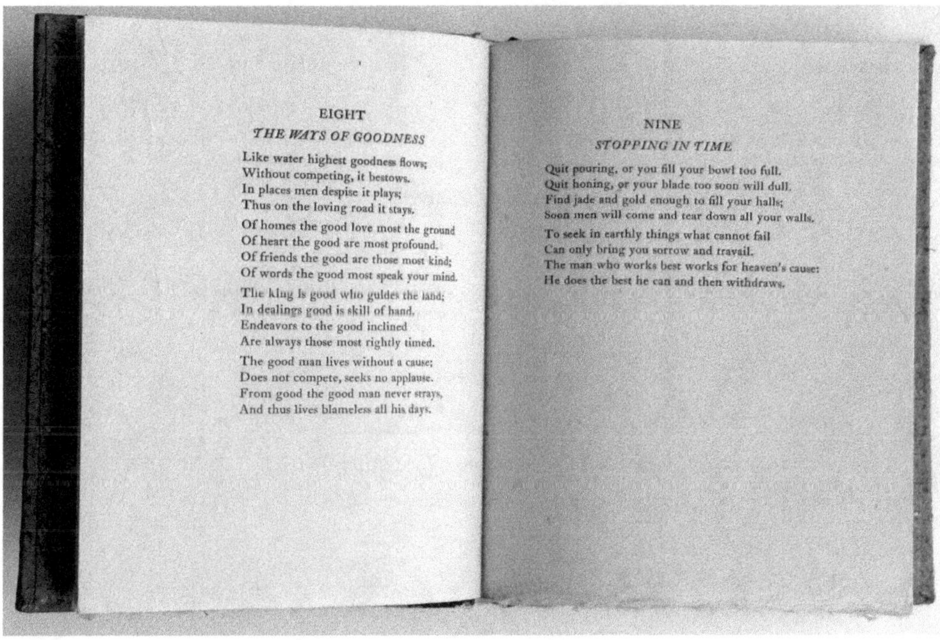

Figure 8- Inside Pages, all handset and hand printed. (c) Chuck Pettis

When I returned to Ithaca, New York, after grad school, I wanted a social service job and became the supervisor of the mail room at Challenge Industries, a sheltered workshop for physically and mentally handicapped people. I led a short noon-time meditation with several of the workers. This was an excellent break from work for me and, I think, also for the benefit of those who joined me. The clients at Challenge Industries were wonderful, and I loved being with them.

Meanwhile, my spiritual studies continued. In 1976, I set up Bliss Press in the garage and hand-printed *Cosmic Geometry: Tantric Transformations*. In the introduction, I wrote:

> "The objective of this book is to present a conceptual basis for the utilization of symbols to evoke positive, sympathetic vibrations in the Self and the Society. This book offers a theoretical foundation for environmental design by the application of the Tantric Principles of Transformation."

By researching and writing *Cosmic Geometry* I realized that I could (and wanted to) create human environments sympathetic to the principles of consciousness that could generate deep psychological impressions upon the individual and accelerate movement to the highest and subtlest state of being.

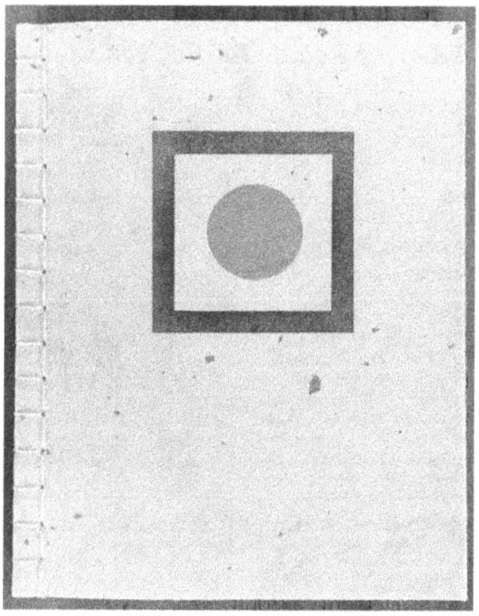

Figure 9- Cover graphic. (c) Chuck Pettis

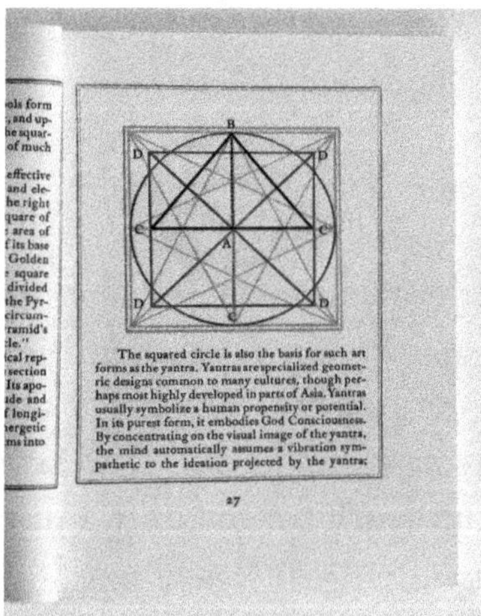

The squared circle is also the basis for such art forms as the yantra. Yantras are specialized geometric designs common to many cultures, though perhaps most highly developed in parts of Asia. Yantras usually symbolize a human propensity or potential. In its purest form, it embodies God Consciousness. By concentrating on the visual image of the yantra, the mind automatically assumes a vibration sympathetic to the ideation projected by the yantra;

27

Figure 10- Sample inside page, all handset and printed. (c) Chuck Pettis

John and Kate Payne – Master Spiritual Healers and Dowsers

My next spiritual teachers were John and Kate Payne in Ithaca, NY. Founders of the Foundation of Light, a center for meditation and healing, John and Kate were extraordinary people with eclectic and diverse spiritual interests and talents. Kate and John traveled widely in the 1960s to gather knowledge about sacred places and seek leading experts in esoteric fields of study. I was hungry for this knowledge and became their apprentice.

Kate was a former Major in the U.S. Air Force with an Interior Design diploma from the NY School of Interior Design and another in professional hat making in Paris, France. She studied French cooking at le Cordon Bleu and became a member of the British Society of Herbalists.

Kate was an expert dowser and taught me and many others the basics of pendulum dowsing. She would give everyone a metal nut with a thread attached as a simple pendulum. She had us sit on a chair, place the thread, and nut pendulum between our legs, and get it swinging back and forth. This was "neutral." Then she had us move the pendulum over our right leg with the intention that our right leg is a "yes" and for us to note which direction the pendulum spins. We did the same with the left leg to get our "no." I was gifted, and I learned quickly.

Kate was an extraordinary healer. She invented the "Growler" system of long-distance healing that uses pendulum-selected sound frequency recipes to heal ailments and diseases. Kate built and maintained a library of over 6,000 volumes on spirituality, healing, and practical arts. With a degree in library science, Major Payne modified the Dewey Decimal System to suit her library's unique holdings.

Colonel John D. Payne was no less unique. John was a 29-year veteran of the U.S. Air Force. After retiring from the Air Force, he accepted a post at the Cornell Center for Radiophysics and Space Research. He finished his professional service as a Professor of Air Science at Cornell University.

John was an expert dowser and healer. A quiet, unassuming man, he was adept at environmental healing and the clearing of negative influences from earth energies and discarnate entities, i.e., ghosts. Along with Terry Ross, John is credited with discovering the mechanics of ley line power centers that were to be helpful to me for the rest of my life.

Figure 11- Kate and John Payne. (c) Chuck Pettis

The Rollright Stone Circle Changes My Life

In 1975, I spent a week in Southern England visiting ancient monuments, including Stonehenge and Avebury. The first place I visited was the Rollright Stone Circle near Oxford. At the time, the Rollright Stone Circle was a center for metaphysical research into ley lines and Earth Energies. When I entered the stone circle, I could feel my consciousness change. When I walked to its center, sat down, and meditated, I found I could go deeper and faster. Other people had similar experiences here. Even dogs could feel the power of this place! During this visit, a couple brought their dog to the stone circle. The dog wouldn't go past the ring of stones. The owners had to take their dog back to their car.

This was my first introduction to places of power. Something was happening here, and I wanted to replicate it. I needed the tools and mind training to investigate their nature. So, with a bachelor's degree in psychology and a master's degree in design, I began a life quest to discover how carefully designed sacred spaces like this could evoke profound psychological experiences. The result was my book, *Secrets of Sacred Space: Discover and Create Places of Power*, available on Amazon.com (https://a.co/d/aT849Np).

Figure 12- Secrets of Sacred Space: Discover and Create Places of Power by Chuck Pettis

There was a lot of interest in earth energies and mysteries in the late 1960s and early 1970s and I bought every book I could find on the subject. Over the years, my sacred space library has expanded to include over 300 books, many out of print, about sacred space, from dowsing, earth energies, geometry, architecture, and astronomic alignments to devas and angels.

After my visits to ancient monuments in England, I learned about a 1969 exhibit on Earth Art with exhibits from Robert Morris, Robert Smithson, and others at the Andrew Dickson White Museum of Art at Cornell University. I was inspired by these artists and my life-long interest in land art, ancient monuments, and sacred space became a life-long pursuit.

Learning How to Dowse

From 1972 to 1995, the Foundation of Light was a fountainhead of healing, metaphysics, and dowsing. Kate, John, and Mabel Beggs, co-founders of the Foundation of Light, brought scholars and savants from around the world to Ithaca which nurtured my curiosity and built my desire to create places of power for spiritual good.

There were many talks and workshops at the Foundation of Light from those special times on avant-garde subjects, including the "Sound of Healing," "A Psychic Look at the Afterlife," "The Spirit of Findhorn," "Communication with the Devas workshop," "Power Centers at Glastonbury, Iona, and Findhorn," and "Map Dowsing and Healing."

At this time, my research on ancient monuments indicated that one of the key components of a sacred place of power is underground water and ley lines. Guy Underwood's *The Pattern of the Past* (https://a.co/d/gwKGGYT) showed that underground water lines were associated with sacred sites. Alfred Watkins' *The Old Straight Track* (https://a.co/d/d1YdofQ) showed that ancient sites in England were aligned in straight lines – ley lines. John Michell's *View Over Atlantis* (https://a.co/d/fgWTNkN) romanticized ley lines and earth mysteries.

John Payne worked one-on-one with me, teaching me how to detect underground water, first by dowsing their water well and other known water wells. I began dowsing for water wells in the Ithaca area. In the beginning, I used a dowsing rod and walked the client's property dowsing for water lines. When I found a water line, I would follow it, marking a little map of the property. I would look for the best place for a well and then dowse for the gallons/minute and the depth of the water line. Dowsing while walking around a property took a lot of time so I learned how to dowse a map. Then I would go to the property in person to mark the "drill" spot for the well driller with a stake.

Everyone who dug a well where I had left a marker found water, at the depth and gallons per minute that I predicted. I remember one instance where I found a successful well location 30-some feet away from a dry well that had been drilled previously. There was only one "failure;" the well driller didn't like my location and drilled elsewhere.

When you dowse for a water well, it is confirmed by drilling. The water is either there or not. Dowsing for ley lines is more subtle than dowsing for water lines; there is no physical form of proof.

I learned about ley lines experientially by visiting many sacred spaces, including the Avebury stone circle, West Kennet Long Barrow, and many more. I walked around until I felt a field of energetic Yang energy emanating from the center of the stone circle (Rollright Stone Circle) or chambered mound (West Kennett Long Barrow). I learned by sensitizing myself to the sensations of ley line and water line earth energies until I could reliably feel and identify them.

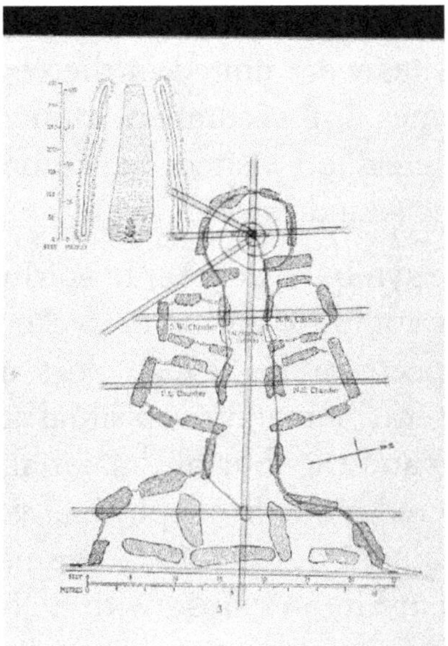

Figure 13 - Ley Line map of the West Kennett Long Barrow.

Creating the Ellis Hollow Stone Circle

I wanted to create a stone circle with the same power I felt at stone circles and mounds in England and Scotland. It was time to put my research into practice. In 1977, I designed and built the Ellis Hollow Stone Circle with help from John, Kate, my sculptor friend Mike Sweeney, and the Foundation of Light Cosmic Monument Study Group.

Figure 14- Summer Solstice Sunrise. (c) Chuck Pettis

On the Foundation of Light property off of Turkey Hill Road in Ithaca, NY, I located two water springs, one ley-line power center, and four ley lines. To create this sacred space with the symbolism of wholeness, harmony, and spiritual union, a recumbent stone and a tall standing stone were placed seven feet apart. Opposites were united in the center of the stone circle. I had seen standing stones and recumbent stones placed close together in Scotland. In one stone circle I visited, the recumbent stone became an acoustic reflector evoking sound waves.

Seven, the Heptad, is the symbolic number of spirituality. The heptad symbolizes perfect order – a complete period or cycle. The musical octave has seven notes, the color spectrum has seven colors, and the week has seven days. The heptad represents humanity's spiritual and mystical nature, our threefold spiritual body, and the fourfold material body. These attributes make the heptad the sacred number of spirituality and religion. With this amazing symbolism in mind, I put 77 - 1977 pennies under the recumbent stone. The outer stones are in an ellipse with a circumference of 77 feet, oriented North-South.

In keeping with the alignments of the English and Scottish stone circles I had visited an outlier stone was linked with the standing stone to mark the Summer Solstice sunrise. It worked! Success! Not only did the alignment to the Summer Solstice sunrise work, but the stone circle had the "feel" of other stone circles of power I visited in England. We had discovered how to create places of power. The Ellis Hollow Stone Circle is still humming away almost 50 years later, and I hope it will continue for hundreds of years.

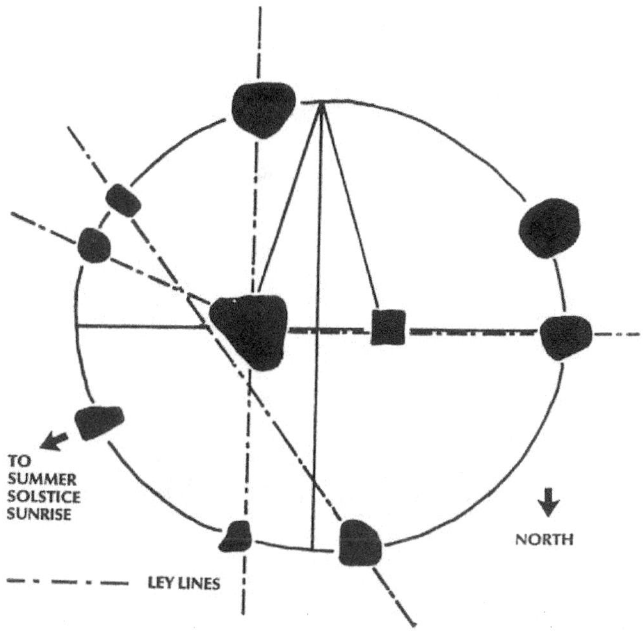

Figure 15- The Design of the Ellis Hollow Stone Circle. The triangular stone in the center is the recumbent stone with 77-1977 pennies under it. The square stone in the middle is the standing stone.

Dorothy MacLean Teaches Me to Communicate with Devas

I met the larger-than-life Dorothy MacLean of Findhorn at the Foundation of Light. She taught me how to communicate with Devas and the spiritual hierarchy. Dorothy Maclean was one of the three founders of the Findhorn Foundation, a spiritual community in northeast Scotland. Dorothy was renowned for her ability to communicate with Devas—nature consciousnesses embodying creative intelligence. She demonstrated this through extraordinary feats, like growing 40-pound cabbages in Scottish sand where vegetables like this had never thrived before.

Dorothy visited Earth Sanctuary with me around 2010. She lived in Issaquah near Seattle for many years with David Spangler, an American spiritual philosopher and pioneer in the New Age movement. Together, they helped transform the Findhorn Foundation in northern Scotland into a center of residential spiritual education. Sadly, Dorothy died in 2020 just after her 100-year birthday.

Figure 16- To learn more about Dorothy and her work with Devas, I recommend the book, "To Hear the Angels Sing: An Odyssey of Co-Creation with the Devic Kingdoms. (https://a.co/d/0f6yGZB)"

Discovering the Energies of Water Lines and Ley Lines

Not all places of power are filled with love and light. Early in my explorations of ancient monuments, I noticed that some felt dangerous or scary. I learned from John and Kate Payne that water lines and ley lines can affect people.

Water lines, when polluted, can cause geopathic stress – negative Earth energies that can harm health, and lead to diseases like cancer. John and Kate taught me how to do a geopathic survey (geo-Earth, pathic-disease) to neutralize these energies and restore harmony.

For example, I once dowsed a friend's home after he complained of feeling unusually sluggish at work. By mapping his house, I found a water line under his desk. After neutralizing it, his energy levels returned to normal.

It wasn't long before I found some homes and ancient monuments that also had ghosts or spirits.

Clearing Ghosts and Spirits

The problem of ghosts or spirits required research. To help me in my research, I read Raymond Moody's book, *Life After Life* (https://a.co/d/8KCTmHk) and Elisabeth Kubler-Ross' book *On Life after Death* (*https://a.co/d/0HR-*

w4C9). These researchers studied thousands of near-death experiences, i.e., people who clinically died, but then came back to life to recount their experience. If you would like to read more about what happens when you die, I recommend *Mind Beyond Death* (https://a.co/d/g7S8bTU) By Dzogchen Ponlop and *Peaceful Death, Joyful Rebirth* (https://a.co/d/7HGUIVj) By Tulku Thondup.

From this research and my teachers, John and Kate, I learned that when we die, we leave our body. When you leave your body, you can see, but you can't be seen. You can hear, but you can't be heard. Then we enter a dark "tunnel" at the end of which is a bright, luminous, loving, and spiritual light. Advice: when you die, do not remain here. Move on. Do not think of your past life and what you are "losing." Be completely unattached to your past life, including your loved ones. Be in the present. There is no going back.

From John and Kate, I learned that some people can't make the passage to the spiritual light and after dying end up stuck here in our world as discarnate humans (ghosts and spirits). They can see and hear, but they can't eat. Because of that in Tibetan Buddhism, they are called hungry ghosts.

Signs of a Ghost's Presence

- A place in a home or space that feels "off," "scary," or even malevolent
- Residents experience drained energy or unexplained fatigue.
- A sensation of being watched despite no visible presence. A young friend recently felt that she was being uncomfortably hugged in a church by a spirit. I map-dowsed the church and found four ghosts. I passed them on, and comfort returned.
- Pets act strangely, barking or staring at empty places.

Since the mid-1970s, I've been clearing people's homes and property by map dowsing. The client mails me a sketch of their home or property and describes their concerns. Then with my pendulum and pen, I dowse the map. If negative water lines or ley lines are found, I neutralize them with a simple spiritual practice and steady focus. If ghosts are present, they are passed over.

I have cleared hundreds of homes and places, primarily through map dowsing. Clients send me sketches of their homes, and I use a pendulum to identify and neutralize negative energy lines and help spirits pass over. For more information, visit https://earthsanctuary.org/geopathic-survey/.

Benefits of a geopathic survey include:

- Eliminating the influences of entities and ghosts.
- Neutralizing negative Earth energies that may be causing ill health and disease.
- Identifying optimal locations for sacred spaces, gardens, and environmental artworks.
- Providing peace of mind and contributing to planetary healing.

When I do a geopathic survey, if ghosts are present, they are passed over into a stage called the "bardo," the intermediate stage one experiences after dying and being born again. I do this special spiritual practice with the knowledge and blessings of my spiritual teachers.

While I am on the subject of ghosts and spirits, I should mention that I can feel them, and they can feel me; they know that I know they are there and appear to me non-physically. When I first visited NewGrange, a very special chambered mound, in Ireland in the late 1980s, I took the tour into the central chamber. I immediately felt hundreds of ancient human entities there and passed them all over. They were happy to be released and "set free." The same experience happened in Maes Howe on the Orkney Islands, my favorite chambered mound.

People sometimes ask me, "Can you tell me more about who they are?" I consciously don't "get to know" the ghosts. I am not a psychic in that regard. My spiritual job is to relieve the ghosts of suffering and enable them to get on with their existence. I get in, do my job, and get out. This keeps everything clean and clear and also keeps me out of trouble!

Learning From His Holiness Jigdal Dagchen Sakya

In 1995, my wife, Claudia, introduced me to His Holiness Jigdal Dagchen Sakya (Dagchen Rinpoche), Head Lama of Sakya Monastery of Tibetan Buddhism. This was a major turning point in my spiritual journey. Tibetan histories indicate that the Sakya family is descended from celestial beings. H.H. Jigdal Dagchen Sakya was born in 1929 in Sakya, Tibet, and was raised and trained to assume the role of "throne holder" of the Sakya school, a role third only to H.H. the Dalai Lama in rank.

I was blessed to receive profound spiritual teachings and initiations from him before his passing in 2016. Here is a YouTube video (https://shorturl.

at/QxvOg) with me talking about Tibetan Buddhism and how it plays a big part in my life.

Pilgrimage Reflections: Sacred Buddhist Sites and Their Spiritual Power

My wife, Claudia, and I went on two pilgrimages to the holy Buddha sites in India and Nepal, once in 1998 and again in 2003. A pilgrimage is a purposeful spiritual journey to sacred places as an act of devotion or respect. In Buddhism, a pilgrimage to the holy places of Buddha's life can erase the negative karmic residue of a lifetime or multiple lifetimes. A pilgrimage to the holy Buddha sites is a never-to-be-forgotten spiritual experience.

At each pilgrimage site, H.H. Jigdal Dagchen Sakya led us in powerful Tibetan Buddhist meditations. Each location's spiritual vibration has been enhanced by thousands of pilgrims praying and meditating there over hundreds of years.

- Lumbini: The birthplace of Buddha
- Namdak Stupa: Where Buddha ordained himself
- Nairanjana River and the Mahakala Cave: Where Buddha meditated to achieve the highest stage of meditation
- Bodh Gaya: The site where the Buddha attained enlightenment
- Sarnath: Where Buddha taught his first sermon after attaining enlightenment at Bodh Gaya
- Varanasi: Where Buddha turned the Wheel of Dharma
- Vulture Peak Mountain: Buddha's favorite retreat in Rajagaha and the scene for many of his discourses, including The Heart Sutra, the essential teaching on emptiness
- Shravasti: Where Buddha displayed great miracles and showed his supernatural abilities
- Kushinagara: Where Buddha passed into Nirvana

Some of us are drawn to mountain tops, some to caves, others to forest clearings. For each of us, there is a place of alignment. Pilgrimage has no further purpose. Being there is reason enough[1].

Figure 17 - Reclining Buddha in the Nirvana Temple at Kushinagara. (c) Chuck Pettis

Arrival at Kushinagara: Honoring Buddha's Parinirvana

In October 1998, after a chaotic day of crowded train rides where we secured our luggage with cables and locks to prevent theft, we arrived at Kushinagara. This sacred site marks the place where Buddha passed away. Within the Parinirvana Stupa, a statue of the reclining Buddha invites visitors to reflect on impermanence. Here we lit candles and chanted sacred Tibetan Buddhist texts and mantras: *The Aspiration of Samantabhadra, The Heart Sutra,* and *The Three-Part Sutra: Confession Before the Thirty-Five Buddhas.* These spiritual practices were led by our esteemed teachers, H.H. Jigdal Dagchen Sakya and H.E. Dagmo Kusho Sakya.

Local monks joined us in the temple. While initially quiet, their chanting soon harmonized with ours, creating an atmosphere of unity. The heat was intense, sweat dripping from my forehead as I joined the monks beside me chanting "Om Mani Padme Hung," a mantra radiating love and kindness. Even now, closing my eyes, I can transport myself back to the spiritual magic of this holy place. This experience ignited a desire within me: to create places that evoke profound spiritual connection and transformation.

Figure 18- The Namdak stupa at Sarnath, where Buddha first taught. (c) Chuck Pettis
Power places are seats of the Gods, focal points of divine energy.
They are the source of spiritual revitalization and renewed psychic energy[2].

The Sarnath Stupa: Where Buddha Gave His First Sermon

From Kushinagara, we traveled by bus to Varanasi near the Sarnath stupa, one of the most important architectural marvels of ancient India and the site of Buddha's first sermon after achieving enlightenment. The nearby Ganges River, sacred to many is steeped in ancient tradition, including cremations along its shores at the burning Ghats. These fires originate from an eternal flame, a poignant symbol of continuity.

At sunrise, our group boarded a long-oared boat to float on the Ganges. Beside us, leaves bearing candles drifted silently on the river. H.H. Jigdal Dagchen Sakya brought his brother's ashes to these sacred waters, a deeply moving act amidst the juxtaposition of death, poverty, and the river's sanctity.

The Mahabodhi Temple: Enlightenment's Sacred Ground

Our journey continued to the Mahabodhi Temple in Bodh Gaya, where Buddha attained enlightenment. This nine-storied pyramid, one of the most extraordinary sacred spaces on Earth, marks the moment Prince Siddhartha became Buddha, forever shaping human thought and spirituality.

Figure 19- The Mahabodhi Temple at Bodh Gaya. (c) Chuck Pettis

Figure 20- Monks doing prostrations in front of the Mahabodhi Temple. (c) Chuck Pettis

At the temple entrance, monks performed prostrations, rising and falling continuously in prayer. Rinpoche explained that prostrations help remove physical obscurations, clearing the path to perceiving one's higher self and achieving enlightenment.

Figure 21- This circular black stone just inside the temple marks the ley-line power center. (c) Chuck Pettis

Point Zero

We entered a foyer with a large Buddha statue and a circular black stone. A monk explained, "This black stone is 'point zero.'" He said this is the only place on the planet stable enough to withstand the energy of Buddha's enlightenment. This black stone marks the most powerful ley-line power center on the planet. This holy power center is amplified by the tens of thousands of people and pilgrims who visit here and do prostrations, meditations, and prayers.

Figure 22- Monks circumambulating the Temple. (c) Chuck Pettis

Figure 23- H.H. Jigdal Dagchen Sakya entering the Bodhi Tree sanctuary directly behind the Mahabodhi Temple where we meditated. This Bodhi Tree is a descendant of the Bodhi Tree under which the Buddha gained enlightenment. (c) Chuck Pettis

Figure 24- H.E Dagmo Kusho (Dagmola), H.H. Jigdal Dagchen Sakya's wife, in front of a Tara statue on the outside wall of the Mahabodhi Temple, adorned by katas, white scarves. (c) Chuck Pettis

The Mahakala Caves: A Testament to Austerity

Nearby, we visited the Mahakala Caves, crossing streams to reach a remote village. When we arrived, our guides, after buying all the cookies and crackers in the previous villages, began giving them to the local villagers while we walked up the trail to the caves.

Figure 25- A meditation room at the Mahakala Caves monastery, near the Mahabodhi Temple where we lit candles. (c) Chuck Pettis

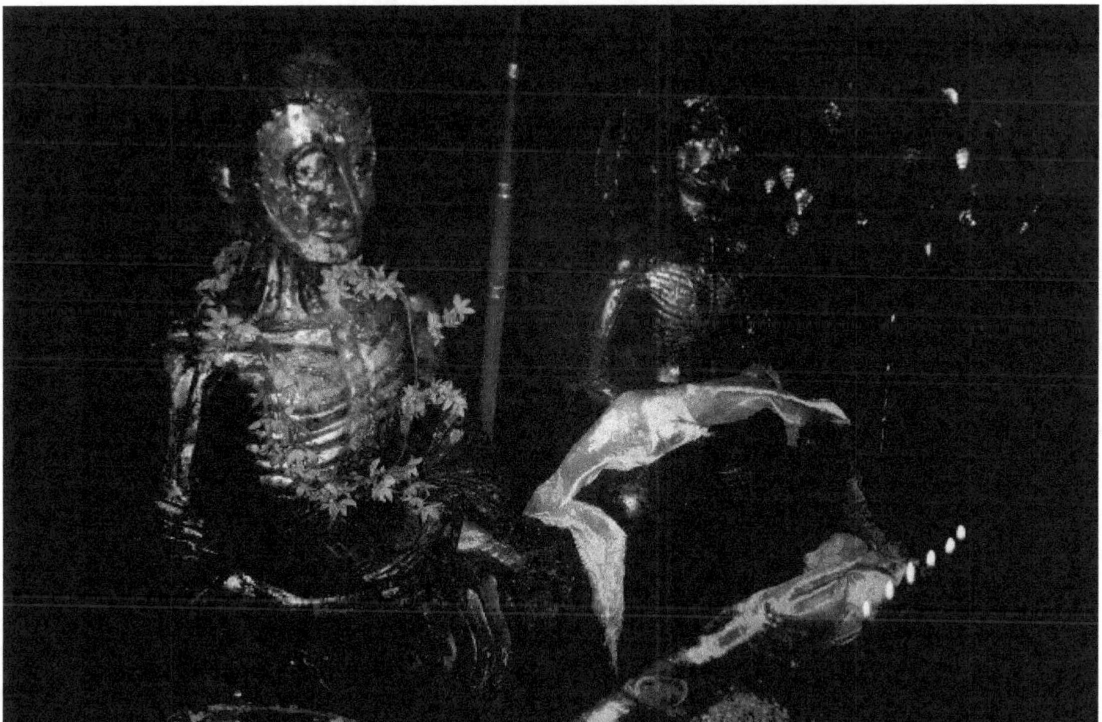

Figure 26- Inside the Mahakala Cave. (c) Chuck Pettis

I will never forget meditating in the Mahakala Cave with Claudia. It was very hot, and the sweat dripped off our faces in this dim, candle-lit cave. The statues in the cave depicted Buddha's time on the Niranjana River where he fasted and performed austerities. If you look at the photo above, you might get a sense of the feeling in this cave and imagine the emotions of all those who had been in this cave. A place never to be forgotten.

Vulture Heap Mountain: Where Buddha Taught The Heart Sutra

Another sacred site we visited was Vulture Peak Mountain, where Buddha taught *The Heart Sutra.* This profound text encapsulates the essence of his teachings.

Teyatha Om Gate Gate Paragate Parasamgate Bodhi Soha
"In this way: OM gone, gone, gone beyond,
Gone completely beyond, enlightenment. So be it!"

This mantra is a daily recitation for me, embodying the aspiration to transcend suffering and attain enlightenment.

Reflections on Pilgrimage

Our pilgrimage brought together living rituals and sacred places, fostering a deep connection to Tibetan Buddhist teachings. The joy of sharing these experiences remains with me, reinforcing the importance of preserving these sacred traditions. Pilgrimages, regardless of one's spiritual tradition, are transformative journeys that deepen faith and understanding.

I encourage everyone to do a pilgrimage of their spiritual or religious tradition to deepen their faith.

Lamdre: A Complete Path to Enlightenment

"There is no highway to enlightenment, only tracks."

In 1999, I traveled to Darjeeling for the profound Lamdre teachings, the highest teachings of the Sakya school of Tibetan Buddhism. This month-long program at the Sakya Monastery in Ghoom provided a comprehensive guide to enlightenment, combining meditation, initiations, and teachings passed through a lineage over 1,000 years.

A stream of enabling circumstances eliminated all excuses. An Internet stock payoff put me in the position for the "Universe" to ask me: "So now you've got money, now what?" The encouragement and support of my wife, Claudia, was key. As she told my curious parents, "He's a nicer person when he meditates."

According to Tibetan Buddhism and many other religions, there is a vast hierarchy of spiritual and enlightened beings whose job is to help all beings attain enlightenment. The walls inside the Monastery are covered with faded paintings and Thangkas (painted fabric artworks) of these Buddhas and Deities.

Figure 27- Painting inside the Ghoom Monastery. (c) Chuck Pettis

The purpose of enlightenment is to change our everyday perceptions into the way enlightened beings perceive the world. Enlightenment is about altering consciousness, the mind, so it can perceive things differently and, most importantly, improve the quality and virtuousness of our actions, speech, and thoughts.

I judge my progress along this path by observing: have I become a kinder, more forgiving, wiser, more patient, and calm person?

Lamdre: A Complete Tibetan Buddhist Path to Enlightenment

Lamdre is presented as a complete path to enlightenment, a remedy for suffering and unhappiness. The two Lamdre texts are *The Beautiful Ornament of The Three Visions* and *The Triple Tantra*. These are difficult texts. Instead,

I recommend The Three Levels of Spiritual Perception: A Commentary on The Three Visions (https://a.co/d/8QKotL6) by Deshung Rinpoche.

Figure 28- Each day began at 8:00 am with the blowing of Tibetan horns, signaling the approach of our teacher, Rinpoche, and his entourage. (c) Chuck Pettis

In the Monastery, packed with several hundred people, the monks would begin each morning with chanting. Imagine being in an ancient monastery and feeling the room fill up with Buddhas and Deities. The hair on my arms goes up now that I remember that time. Powerful internal feelings arose. It is hard to put those feelings into words. I imagine being in a Celestial Heaven, a Buddha realm filled with enlightened beings, Buddhas, Bodhisattvas, and Deities. I am blessed.

Lamdre includes a series of initiations. In an initiation, you obtain blessings and empowerments as part of a spiritual ceremony. The central Lamdre initiation is the Hevajra Sadhana (meditation). The initiation begins with the creation of a sand mandala.

Figure 29- The monks begin filling in the mandala design starting from the center and moving outwards. (c) Chuck Pettis

Figure 30- Almost every day there was either an initiation, empowerment, or special blessing. Here, Rinpoche is bestowing the Long-Life empowerment attended by several thousand people. (c) Chuck Pettis

Figure 31- Chuck Pettis making an offering to H.H. Jigdal Dachen Sakya. A great blessing! Unknown photographer.

My Daily Meditations and Spiritual Practices

I have been blessed to receive blessings, initiations, and Buddha's spiritual practices from some of the world's greatest Tibetan lamas and teachers: His Holiness the Dalai Lama, His Holiness the Sakya Trichen, His Holiness Jigdal Dagchen Sakya, Her Eminence Dagmo Kusho Sakya, His Eminence Avikrita Sakya, His Eminence Asanga Rinpoche, and more.

At Earth Sanctuary, I integrate the teachings I've received into daily walking meditations. Visualizing and praying to Buddhas and Bodhisattvas, I infuse the land and sacred spaces with blessings and the intention that they bless and inspire all visitors and living beings. Visitors are encouraged to meditate and pray, amplifying the sanctuary's spiritual energy through their presence and practice.

The Magic of Earth Sanctuary

Earth Sanctuary invites visitors to experience peace, inspiration, and spiritual upliftment. Through meditation and prayer, each visitor contributes to the sacred energy of the space, creating a ripple effect of blessings and transformation. As we uplift ourselves, we uplift others, embodying the interconnectedness of all beings.

May all who visit Earth Sanctuary find serenity, spiritual growth, and the magic of sacred ecology. Together, we can create spaces that inspire and nurture the highest aspirations of the human spirit.

> Short Instructional Meditation Video for the Earth Sanctuary Dolmen. (https://shorturl.at/7lc3V)
>
> Short Instructional Meditation Video for the Earth Sanctuary Labyrinth. https://shorturl.at/GuKnJ
>
> Short Instructional Meditation Video for the Cottonwood Stone Circle. (https://shorturl.at/6L0f1)
>
> Short Instructional Meditation Video for the Earth Sanctuary Ley Line Sculpture. https://shorturl.at/WVkXL

In this chapter, we have covered the importance of meditation and provided instructions for a variety of calm-abiding meditations. You have learned about my spiritual path and some of the spiritual practices I do to create sacred spaces and sacred ecology. If you are already a meditator, I hope you will grow your meditation practice and apply your insights to creating spaces around you that are uplifting and beneficial to you, your family, and all the plants, trees, animals, and spiritual beings.

"May all beings be happy! May all beings be free of suffering!"

Questions for Group Discussions

These questions should help stimulate a rich discussion on the various aspects of meditation and spiritual practices and their implications on spirituality, ecology, and personal growth.

- What role does meditation play in the author's creation of Earth Sanctuary?
- Do you meditate? If so, how has it affected your life?
- What events in the author's life do you think were most significant?

- Have you gone on a pilgrimage? Where did you go and what did it mean to you?
- Have you ever done a retreat? Where did you go and what did it mean to you?
- What lessons can readers take from the author's journey to apply in their spiritual practices?

Chapter 2:
The Power of Ley Lines and Ley-Line Power Centers

"It's truly remarkable how the ley lines make such a difference in the energy vibrating in certain areas."

The most potent ancient monuments and sacred spaces around the world have one thing in common: the presence of Earth energies (i.e., underground water, ley lines, and ley-line power centers), which have the power to alter and uplift human consciousness. Earth Sanctuary follows this tradition. Its sacred spaces are located over ley-line power centers and are networked with ley lines. Dowsing is the intuitional practice or technique for locating these Earth energies.

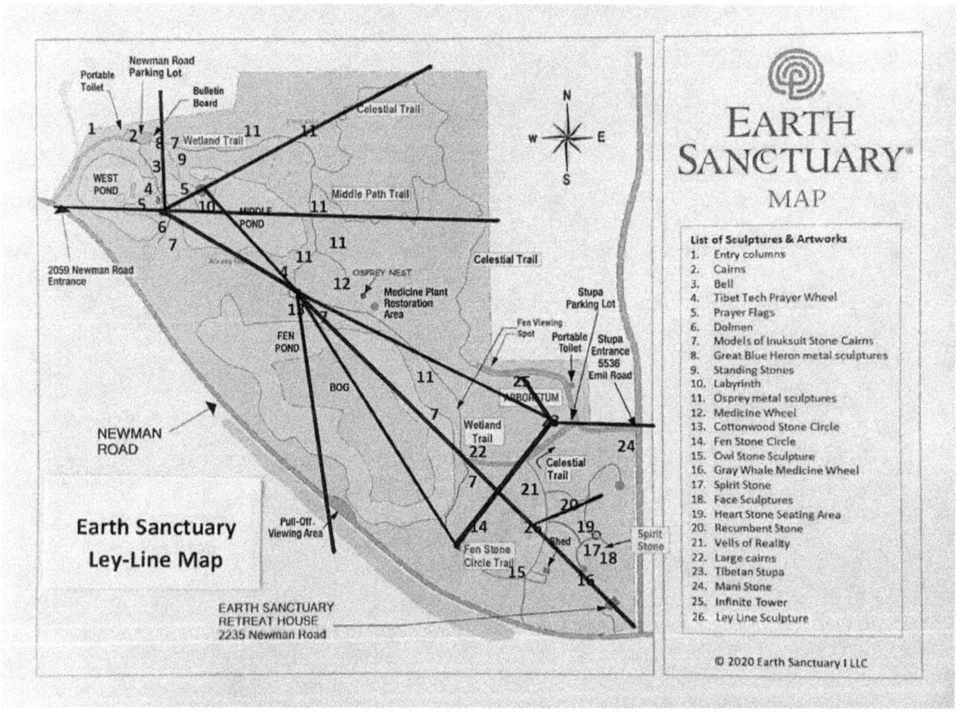

Figure 32- The Earth Sanctuary Ley Line Map (c) Chuck Pettis

Introduction to Ley Lines and Ley Line Power Centers

Everything on Earth is impermanent, flowing, growing, blowing, falling, and rising. Many changes, such as rivers, wind, weather, and tides, are easily seen. Some are invisible to us because our senses can pick up only limited ranges of vibrations and radiations. For example, we can see color but not X-rays. We can hear sounds, but only if they are within our hearing range.

Many of these invisible spectrums can be detected and "seen" by extending our senses via various technologies: radios, smartphones, television, films, amplifiers, readouts, scopes, transducers, etc. As technologies to detect them are developed, new spectrums will be discovered. Until technologies are developed to detect the presence of underground water and Earth energies, we must rely on our body's sensitivity and our intuition to guide us in the right direction.

The human body is the best "receiver" on Earth. We can detect many things that machines and technologies cannot, such as emotions, feelings, and consciousness. Two subtle energy systems on the frontiers of human perception, the electromagnetic fields of underground water streams and ley lines, need to be more recognized, studied, and used today. Let's start with water lines.

Water Lines

This primary water system exists deep inside the Earth as hydrated minerals. At shallower depths, this diffuse "steam-like" field of water gradually coalesces into pockets of liquid water which eventually connect to underground streams. The water travels upward in a vertical shaft called a "pipe" until an obstruction stops its flow. This point is called a dome of water because the water is, in effect, domed up. If fissures or cracks in the Earth are connected to the pipe, the pressure of the water pushes the water into the cracks, which then become what we call underground streams, or water lines, which can then travel for great distances under the Earth's surface. Their course through the ground is generally winding and non-linear.

It is common practice to dig or drill a hole down to the water line to find a water source. Thousands of well drillers are in this business. An artesian well or water spring is formed when a water line flows on its power out from the surface of the Earth.

There is a difference between groundwater and the primary water system described above. Groundwater is part of the above-ground hydrological cycle involving evaporation, cloud formation, rain, rivers, oceans, and the underground water table. Primary water is a below-ground hydrological cycle and therefore is not affected by drought.

Primary water is found at most ancient monuments and temples, and Earth Sanctuary's sacred spaces. Usually, there is a water dome or even a well or a spring at the center of the monument. A water spring, the place where water is available for use from the surface, is a natural spot for building a sacred place. By marking the site, or distinguishing it from other places, the monument becomes "holy ground."

The close relationship of water lines and springs with ancient monuments was established by M. Louis Merle and Reginald Allender Smith in the 1930s. These men were dowsers or diviners of water; they could locate underground streams and springs without using scientific instruments. Merle established that ancient monuments were situated over the crossing of underground streams. Smith went further to say that springs are constantly present at the centers of stone circles and earthworks. This discovery indicated that the selection of sites for ancient monuments was not arbitrary, but a conscious decision based upon the presence of underground water.

Smith's work inspired Guy Underwood, author of *The Pattern of the Past,* to spend many years investigating the connection between ancient monuments and underground water.

Many animals are not only affected by water lines but can instinctively perceive and use them. We, as humans, are also affected, but less naturally, and need artificial assistance to perceive them. Using a pendulum, forked stick, or dowsing rod, when we are over a water line our muscles tense slightly, causing an almost imperceptible reflex movement in the arms and hands, which causes the pendulum or dowsing rod to move and indicate the presence of water.

What Do Water Lines Feel Like?

As water flows through underground streams, it creates a subtle electromagnetic field, several feet wide, that rises vertically above the water line, even through multiple floors and stories. This vertical planar field of electromagnetic energy affects people physically, mentally, and spiritually.

When I ask people what it feels like to stand over a water line, here are a few responses:

- "Slow and warm and fuzzy."

- "Heavy in my arms."
- "Faint undulating energy."
- "Surprise. Electric. Pleasing."
- "A flowing directional pull, like ripples."
- "Chocolate syrup."
- "Calming energy."

It's one thing to experience the effect of a water line by standing over it for a short time. It's another thing to work or sleep over a water line. A water line has a yin, or passive field, associated with it. Being on a water line will tend to slow you down and make you lazy or apathetic. People who work at desks over water lines often have problems getting enough energy to get work done or even to get started.

Water lines can have serious negative effects when the water line is polluted physically or psychically. Negative water lines not only create a place of passivity but can be detrimental to one's physical and mental health. I consider negative water lines to be a major (but generally unknown) cause of disease.

Ley Lines

Along with water lines, ley lines are found at most ancient monuments and sacred places. The honor of the rediscovery of the ley-line system belongs to Alfred Watkins. His basic postulate is that ancient monument sites align in straight lines. Many ancient sites found on British ordinance maps can be connected to form an incredible coincidence of interconnecting lines. A shortcoming of this particular definition of ley lines is that many "ley hunters" have assumed that because three or more sites are aligned, they are a "ley line." Alignment does not determine the presence of an energetic ley line, although it can act as supporting evidence for one.

The ley-line system exists as an independent circuitry that can affect consciousness. Ley lines are part of the Earth's energy system. Sacred spaces and monuments reveal or mark the network, making the sites more special by connecting them.

Ley, as a word, is akin to leoht (light illumination) and Middle English lea meaning "pastureland, a meadow which is open to the sun and therefore, at times, drenched with light." This connection of the word ley with light is significant on several levels. Physically, the clearing of tracks through the forest lights the way and marks the "ley of the land."

The word ley is related to ley, lee, and lay. This etymological sequence describes a sort of cosmic roadway system upon which people traveled in pre-Renaissance times. First, lines were delineated by cleared hilltop notches (ley), then woodland through which the ley line passed was cleared (lay), and then the fields that domesticated the landscape were cleared (lee) with the names ley, lay, and lee applying to each stage of ley landscape development.

Visualize mounded tree groves on ley lines and a grove of trees on the ley lightway, filled with sacred cosmic light. Imagine standing on a hilltop at dusk, seeing an aura of lighted lines passing through earthworks and stone circles, with darkened groves of trees glowing with soft light. A magical mystery tour!

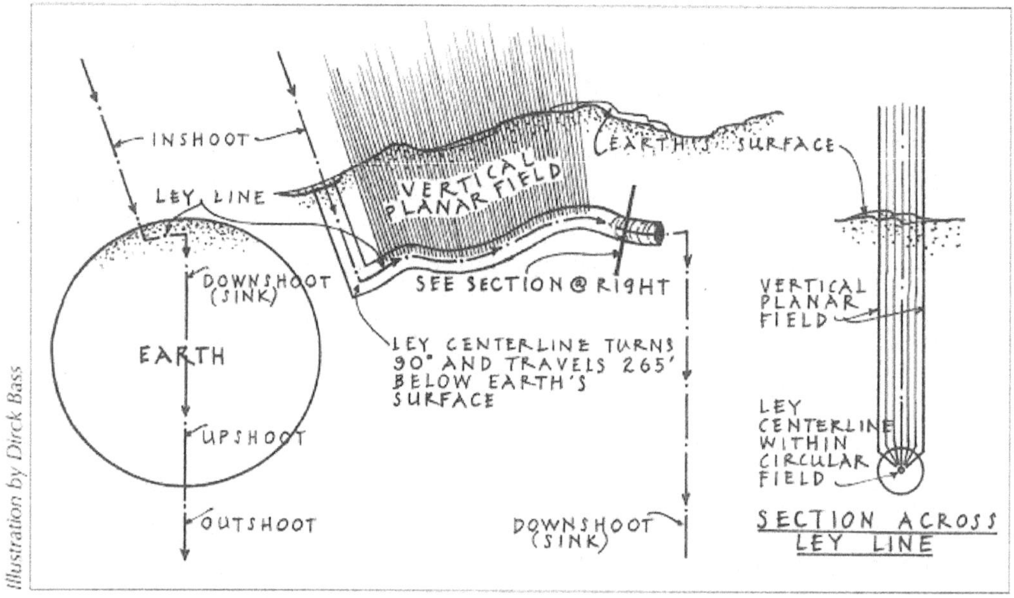

Figure 33 - Ley lines are part of the earth's energy system and exist as an independent circuitry with the capacity to affect consciousness. The nature of this field is yang or energetic.

Ley lines are cosmic forces originating outside of the Earth. They penetrate and leave the Earth vertically at nodes. The penetrating nodes are called power centers. As illustrated above, when entering, ley lines continue to a point 265 feet below the surface of the Earth. At this point, it makes a 90-degree right-angle turn and travels in a perfectly straight line as seen from a "birds-eye view" and in an undulating motion as seen from the side, but always maintaining a depth of 265 feet, relative to the surface of the Earth.

What Do Ley Lines Feel Like?

Like water lines, a vertical field extends up from the ley line through homes and buildings. The nature of this field is yang or energetic. A person who sits or lies over a ley line for an extended time will tend to be hyperactive. This can work to advantage in healing or in situations where extra energy is useful, but if someone is already very energetic, the ley line may cause an unhealthy situation. If the ley line is negative, the negative aspects of extra energy will manifest in tension, anxiety, and neurosis. Here are some comments from people who told me what it feels like standing over a ley line:

- "Energizing, white/fuzziness."
- "Light, see a glowing line that goes with the direction of flow."
- "Began to feel like I was weaving."
- "Faint smooth energy."
- "I felt crackling. Like electricity."

Ley Line Power Centers

At every point where ley lines enter the Earth (inshoot), there is a blind water spring.

A blind spring occurs when underground water rises upwards under pressure and, instead of emerging at the surface like an artesian spring, splits into smaller streams, like the spokes on a bicycle wheel. The ley-line inshoot and accompanying water springs are the universal prerequisites for power centers. It is not just the water spring as suggested by Underwood or the ley line as suggested by Watkins, but the union of the two that determines the site selection of monuments.

Ley lines and water lines have fundamental similarities and differences. They both form a network of force fields over our planet and seem to affect human behavior. Ley lines originate from outside the Earth, while water springs originate from inside the Earth. Ley lines travel in straight tracks with 90-degree turns, while water lines are non-linear and circuitous.

Ley-line power centers are a combination of yang ley line energy and yin water energy. This combination creates a synergetic, holistic field, which is greater and uniquely different from either of the two energies alone. A ley-line power center radiates a universal energy that affects consciousness and can also be enhanced and charged by one's consciousness. This is why I do meditations and my spiritual practices at the Earth Sanctuary sacred spaces. I intend that the blessings of my meditation can spiritually benefit future visitors and all the beings who inhabit Earth Sanctuary.

Ley-line power centers are significant because they can alter and uplift consciousness. I am inspired when I visit ancient monuments and other sacred places. The strong human emotions experienced at a ley-line power center create layers of consciousness over time that future visitors can feel and experience. The field of a ley-line power center attracts and stores strong emotions. For example, if someone has a special spiritual experience, some of that experience is "deposited or saved" at the ley-line power center for future visitors to "withdraw." When you visit Earth Sanctuary, please make your own spiritual "deposits" and "pay it forward." It is much better to pay it forward than to pay it back.

As Earth Sanctuary visitors pray and meditate in one of Earth Sanctuary's sacred spaces with emotion and faith, some energy is "held" by the ley-line power center. If someone subsequently visits that sacred space and needs that energy, they will receive the energy as a blessing. This is called sympathetic vibration. A common example of sympathetic vibration is to sound a tuning fork and bring it close to, but not touching, another fork of the same frequency. The second fork will then begin to vibrate sympathetically.

The power of ancient monument sites and Earth Sanctuary's sacred spaces lies in the interaction of the telluric Earth field of water lines with the cosmic solar field of ley lines. Their combination creates a synergetic, holistic field that is greater than either of the two energies taken separately. This fusion of the fundamental components of the Universe, yang and yin, is the source of all matter, energy, and consciousness.

The power center radiates a universal energy that affects consciousness and can also be influenced and changed by consciousness. As silt becomes sedimentary rock over time, strong human emotions experienced over time at a power center create layers of consciousness that future visitors can feel and experience.

What Does a Ley Line Power Center Feel Like?

- "I felt like I was floating off the ground."
- "Energizing."
- "Light...a glowing line that goes with the direction of flow,"
- "Began to feel like I was weaving."
- "Faint smooth energy."
- "Lightness. A pulling upward."

- "Pulsing, flowing, deep lively sensation just below my solar plexus."
- "Warm and solid."
- "Crisp, clear."
- "Powerful."

This feeling of power as sensed by our consciousness and body is the key thing to seek at any sacred place -- it is the effect of the field on our consciousness that counts, not the name, technicalities, or details.

When you visit ancient monuments or sacred places, be aware of and experience your level of consciousness. Feel your mood change, what kinds of thoughts you have, and what "comes to mind." If you have negative feelings or don't feel a place is safe or "right," avoid it.

Ancient monuments are a blessing because they elegantly mark power centers. In many parts of the world, all you have to do is find a megalithic monument, mound, or ceremonial place and you've found an important power center.

How can you know if a place is on a ley line power center?

A ley line power center has a feeling and a special sensation. If you can visit Earth Sanctuary, walk to the center of each sacred space, close your eyes, and feel your sensations. How does it feel? Remember that sensation. Spend time at the Ley Line Sculpture. Walk in and out of the sculpture until you can sense the difference.

If you are in Seattle, Washington, visit the Sakya Monastery of Tibetan Buddhism. The white monument, Stupa, on the corner is on a ley line power center. If you attend the 10:00 am Sunday meditation at Sakya Monastery, the Shrine is on a ley line power center and the center aisle marks a ley line.

If you are in England, visit the Rollright Stone Circle near Oxford. Most ancient monuments in the United Kingdom are on ley line power centers. In Europe, most ancient cathedrals are on ley line power centers.

As you travel, visit places recognized as sacred and visited for their spiritual power.

What Comes First: The Sacred Space or the Ley Line Power Center?

When I first purchased the Earth Sanctuary property, I dowsed it and found an existing ley line power center on a small hill between the Middle and West Ponds. I had an idea for the design of a Dolmen which was later installed on the ley line power center.

At all the other sacred spaces at Earth Sanctuary, the ley line power center came *after* the optimal location was determined. For example, we designed the labyrinth to be centered on the little peninsula by the Middle Pond. Then, the Landscape Angel and the Spiritual Hierarchy brought in the ley line power center and the connecting network of ley lines.

Just before the base of the Infinite Tower was completed, I put quartz crystals under the tower. Immediately I felt the inshoot of the ley line come down. Cooperating with the Landscape Angel and the Spiritual Hierarchy is one of the keys to making a sacred space powerful. I sense that they are happy to help if we do the work on the physical plane.

Amplifying Power Centers

Ley line power centers can be amplified. When we built the Ellis Hollow Stone Circle in Ithaca, NY, we learned by dowsing and watching our sensations that putting quartz crystals under the stones amplified the energy. The electronic circuits in watches, computers, and medical equipment employ quartz crystals because they have a precise and constant frequency, making them excellent for controlling the frequency of electrical impulses.

I put quartz crystals under all standing stones and Earth Sanctuary sacred spaces. In addition, I place and bury crystals, stones, and minerals that are the opposite energy of the above-ground stones. I credit Collen Marquist, crystal expert and author of the book <u>Crystalline Communion, Mineral Properties for Healing and Integration</u> (*https://a.co/d/fqmudrV*), for determining the best crystals and minerals to go under each sacred space.

Herkimer diamonds and black tourmaline are placed under the Earth Sanctuary Labyrinth's Pennsylvania Blue Stones at strategic locations. Turtleback stones, the most yin stone in the world for this purpose, are placed under the Yang Columbia River Basalt columns in the Cottonwood Stone Circle. This is the design technique of juxtaposed opposites. Placing opposite forms next to one another creates a wholeness, an ordered form, which creates the potential for transcendental experiences, a creation of a new reality.

Dowsing

Dowsing (deuten, to declare, douse, to plunge) refers to techniques for finding water or other things by using a dowsing instrument, such as a dowsing rod. The use of the dowsing rod, a Y-shaped forked stick or rod used to find water and hidden objects, is probably as old as humankind. The dowsing rod is still used commercially by many well-drillers and contractors who have to dig around underground pipes and lines.

Dowsing is searching for anything by projecting an intent of what is desired and receiving confirmation or non-confirmation feedback through the body, usually by the movements of a dowsing instrument. It is a form of clairvoyance, the ability to see at any given moment what is happening elsewhere. Our senses are more powerful than we think. Because our physical and psychological apparatus is designed to satisfy our desires, they realize their potential only to the extent we utilize them.

Divining is a close synonym of dowsing and gets to the root of what dowsing is all about. Divining comes from the word *divinus* meaning "of, or by, or for a god, the gods, also inspired by them." Hence, divining is a spiritual practice -- the success of which rests on a divine state of mind.

There are many books on the history of dowsing, the exploits of successful dowsers, and the many uses and applications of the dowsing technique. Dowsing is a natural tool that enables you to amplify what you perceive but have not bothered to pay attention to before. It is a handle on the abstract world of feeling, intuition, and the sacred.

Dowsing can be used to:

- Locate underground water lines and springs. The direction of flow, depth, and quantity of flow (gallons per minute) can also be ascertained.
- Locate ley lines and their direction of flow.
- Find power centers, places where the Earth's field alters human consciousness.
- Determine if a water line or ley line is harming the health of the people living or working above the line.
- Communicate with and receive guidance from Devas and Angels.
- Map underground earth energies to design and build sacred spaces.

Learn How To Dowse

To get started with dowsing, you must first make, buy, or borrow a dowsing rod, pair of L-rods, or a pendulum. I recommend the pendulum.

A pendulum is simply a weighted object hanging from a string or bead chain. The chain is held between the thumb and forefinger, which should be pointed down, so the fingers look like a long-beaked bird. The pendulum can be held at rest, but more often it is held so that it swings back and forth, which allows a quicker dowsing response than just holding it at rest. The typical pendulum response is to swing clockwise or counterclockwise in a circular motion when you pass over a water or ley line. The clockwise and counterclockwise motions can also be coded to imply opposites like positive or negative, yes or no, yang or yin.

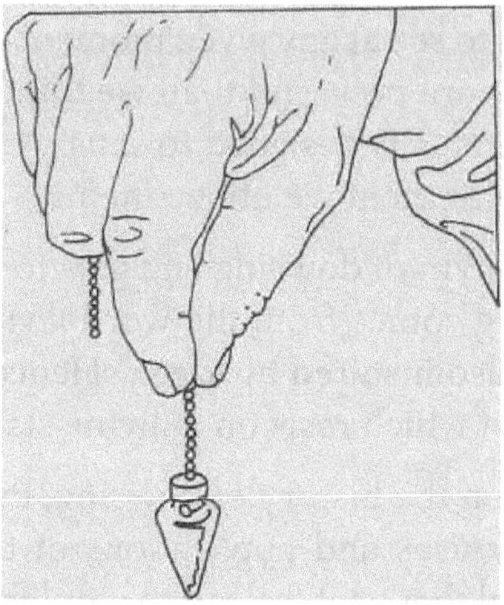

Figure 34 - Hold the pendulum with the thumb and forefinger so it hangs naturally.4

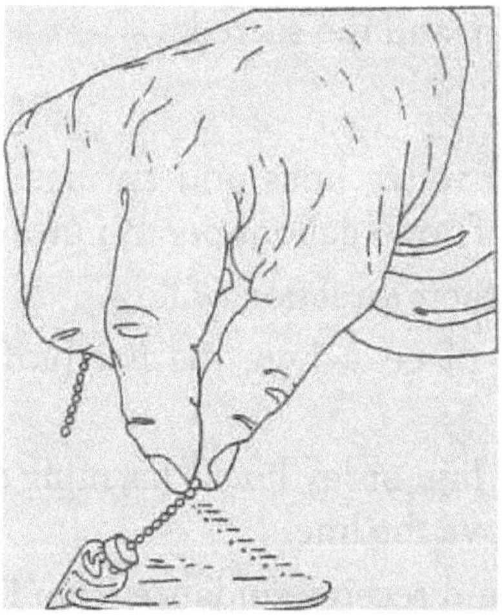

Figure 35 - The pendulum swings in a circle when over an earth energy.5

You can purchase pendulums, use a necklace, or make a pendulum. One of the easiest and quickest ways to make a pendulum is to get a nut (the bolt kind of nut) and tie a thread or light string to it. You can also use a stone, crystal, button, or another small object. You'll get a better response if the object is circular and balanced. Otherwise, the pendulum swing may be slightly off.

The portability of the pendulum makes it a good general-purpose dowsing tool that you can carry in your pocket. It can be used for all types of dowsing. It's best for map dowsing and close detail work.

Coding the Movements of Your Dowsing Instrument

Our bodies are wonderful receivers, capable of picking up many signals. The dowsing instrument doesn't move all by itself; it simply reinforces the body's natural dowsing response, making it more visible and reliable.

The pendulum is the most versatile dowsing instrument and is adaptable for uses other than looking for water or ley lines. For example, you can find the polarity of a water line (positive or negative) or set up a yes/no communication system (a code between you and your body) for yourself. A clockwise spin is usually yes or yang (positive). A counterclockwise spin is usually no or yin (negative). The yes/no correlation with the direction of spin varies from person to person. Therefore, establishing your spin direction for yes and no is the first thing dowsers should learn.

Establishing Your Yes/No Code With the Pendulum.

Sit on a chair, upright, with your feet spaced about a foot and a half apart. Do not let your feet or legs cross. Hold the pendulum between your legs and let it swing back and forth. The back-and-forth swing is like the neutral gear in your car — the engine is running, but the car isn't moving. Next, make the pendulum move in a clockwise circle. Then make it move in a counterclockwise circle. Play with it for a while to get a feel for how the pendulum feels when it's moving in different directions.

To find what yes (yang) is for you, hold the pendulum about two feet above your right knee. Slowly lower the pendulum toward your knee. When the pendulum enters your aura - the electromagnetic field of your body - it will begin to spin. Because the right side of most people's bodies is yang in nature, the direction the pendulum spins is your code for yes. Bring the pendulum back between your legs and let it return to the neutral back-and-forth swing. Then, follow the same process over your left knee to find your code for no (yin). If it's not working, make sure your legs aren't crossed, meditate for a few minutes, take a deep breath, and concentrate on the task.

Check your yes/no code out with places or questions you know the answer to and confirm that your code is consistent with the results you get. The goal is to find a yes/no code that works reliably for you.

Get Used to the Feel of Your Dowsing Instrument

Practice making your pendulum spin. Practice is important because the main problem beginning dowsers have is getting the feel of the dowsing instrument. Remember, the dowsing instrument is not moving magically on its own. You are making it move. Like learning to play a musical instrument, it takes a while to get used to how each dowsing instrument works and reacts.

Dowsing Earth Energies

There are five basic steps for dowsing earth energies:

- Calm your mind and attain a state of mindfulness.
- Concentrate your mind on the desired object.
- Survey the area by walking around the site.
- Mark water lines, ley lines, and ley line power centers.
- Have your results checked.

Biofeedback has shown that we can train ourselves to monitor and regulate our bodily processes, which are usually automatically handled by our body with only minimal conscious awareness. We can train our mind and body to pick up subtle earth energies and indicate their presence through muscle and nerve interactions, resulting in a visible response by the dowsing instrument.

Tuning in to whatever you are looking for is as easy as thinking about what you are looking for. If you are looking for water lines, imagine water flowing underneath the ground in a crack or fissure. If you are looking for ley lines, they are a little more subtle since there isn't a physical substance to detect them. It's an energy you are seeking, yang in nature.

Your state of mind is all-important in dowsing. If you meditate, you can get into the dowsing state of mind automatically. Put your mind in a receptive and open state. If you don't meditate, use your powers of concentration to focus your attention on what you are looking for.

For more on dowsing, see the chapter in my *Secrets of Sacred Space* book on how to dowse or find a local dowser who will teach and show you how to dowse. Another option is to connect with the American Society of Dowsers, the Canadian Society of Dowsers, or the British Society of Dowsers. They have classes and conferences where you can meet other experienced dowsers.

The Earth Sanctuary Ley Line Sculpture

Land art inevitably must not only mark an earth event but also draw attention to the beauty of the landscape to help people feel and see the earth in a new way. I created the Ley Line Sculpture so that people can experience the actual feeling of ley lines and ley line power centers so that they can recognize them when visiting sacred places such as old European cathedrals and ancient monuments.

Figure 36 - Experience the power of ley lines and ley-line power centers at the Ley Line Sculpture. (c) Chuck Pettis

I gathered driftwood from the beach after it had been rolled in the sea and scorched from the sun. I burnt each one with a torch. The burnt driftwood poles were located beside a rounded earth mound to mark the significant ley line running from the Earth Sanctuary Retreat Center to the Cottonwood Stone Circle.

To experience the sensation of ley lines and ley line power centers, visit Earth Sanctuary's Ley Line Sculpture. To experience a ley line, stand between two of the driftwood poles and feel your sensations.

Figure 37 – The diamond-shaped area at the end of the Ley-Line sculpture is a Ley-Line Power Center – (c) Michael Stadler

At the lower end of the line is a ley-line power center marked by a small driftwood diamond-shaped space. To experience a ley line power center, stand or sit in the middle of the diamond-shaped space. Some people see white light, others feel like they are floating. Some start swaying back and forth. What do you feel?

Short Instructional Meditation Video for the Earth Sanctuary Ley Line Sculpture

https://shorturl.at/5xHDQ

The Seattle Ley Line Map

Figure 38- The Seattle Ley Line Map (c) The Geo Group LLC.
Seattle photo courtesy of Cosmic Images, Bellevue, WA.

In 1987, The Geo Group applied for and received a $5,000 grant from the Seattle Arts Commission to "...explore their visions for the city of Seattle ... to consider the fantastic as well as the realizable." The project was commissioned for the Visionary Collaborations portion of Northwest Interpretations to represent what Seattle as a City might become.

At the time, the Geo Group consisted of four architects and artists, including Dennis Alkire, Becca Hanson, Mike Sweeney, and myself. We, The Geo Group, created the Seattle Ley Line Project making Seattle the first city in America - possibly the first city anywhere - to have its ley lines located and mapped.

The purpose of the Seattle Ley-Line Project was to:

- Identify and map the major ley lines and ley line power centers within the City of Seattle using dowsing techniques; and
- Design and build a series of environmental artwork that will be used to mark and enhance the ley-line energy.

To create the Seattle Ley Line map, I map-dowsed a large paper map of the City of Seattle. I map dowse by holding a pen in my left hand and my pendulum in my right hand. I focus on "ley line" as I move the pen over the map. When I feel a ley line, I stop, make a pen mark, and then follow the line, marking it on the map. There is a chapter in my *Secrets of Sacred Space* book on how this was done, and the dangers of dowsing a large city like Seattle.

Dangers of Dowsing a Large City

Map dowsing is like flying over the entire city and probing for places of power. Not all power centers are positive and may be harboring a bee's nest of human entities (i.e., ghosts) and non-human entities (e.g., demons). I have been trained in spiritual practices on how to deal with entities and have been permitted to perform these spiritual practices when needed by my spiritual teachers, most recently, H.H. Jigdal Dagchen Sakya.

After dowsing the Seattle Ley Line map, I went with some friends to find and verify each of the ley line power centers. Most are on private property and not accessible by the public. There was one ley line power center that was very scary. When we stopped outside the house where the center was located, the negative energy was palatable. No one would get out of the car. We could see a person in the second-story window standing with their hands raised touching the window.

If interested, there are several line power centers in Seattle public places: The Washington Park Playfield at the Arboretum, a quite magical one in Interlaken Park, and one in Roanoke Park. Directions are here - scroll down: https://earthsanctuary.org/geopathic-survey/the-seattle-ley-line-map/

After completing the large ley line map, The Geo Group created an artwork showing the Seattle Ley-Line Map and proposed environmental artwork as exhibited as part of Seattle City Light's 1% for Art Portable Works Collection. The artwork is encased in a wood and metal case in the shape of a squared circle. The name "squared circle" comes from circumscribing a circle by the square, and also from the ability to geometrically construct a unique square and circle such that the square's perimeter is equal to the circle's circumference to within 1.82 percent accuracy. The squaring of the circle unites the symbolic image of the square with that of the circle together in a balanced symbolic form. It represents the manifestation of heaven on earth and the reconciliation of the rational (square) and irrational (circle).

In the center of the case is a large map of the Seattle ley lines and on either side of the map is a description of the project and proposed artworks to be built on the power centers in the city.

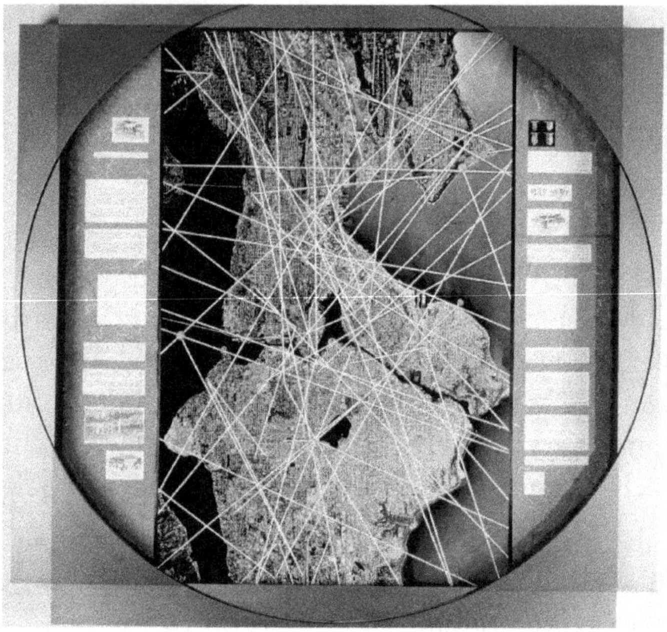

Figure 39 – The Seattle Ley Line Artwork on display in Seattle City Light's 1% for Art Portable Works Collection. (c) Chuck Pettis

In this chapter, we have covered the importance of water lines, ley lines, and ley line power centers to create sacred spaces that have the power to alter and uplift human consciousness. We have explored the concept of ley lines and ley-line power centers, highlighting their significance in enhancing the energy of certain areas and their historical importance. Ley lines are described as part of the Earth's energy system, with the power to influence human consciousness. The presence of these lines at ancient monuments and sacred spaces suggests that their intersection with underground water streams creates powerful energy centers. These power centers amplify consciousness and emotional experiences, thus making them key locations for spiritual practices and meditation.

We also explain the role of dowsing, an intuitive practice used to locate these Earth energies, including water lines and ley lines. The text describes how these subtle energies, though invisible, can be sensed and detected by humans, sometimes with the aid of tools like dowsing rods. Ley lines are characterized by their straight paths and the energizing effects they have on individuals, whereas water lines are associated with a calming energy. The combination of ley lines and water lines at power centers is said to create a synergistic effect, enhancing both spiritual and physical experiences for those who visit these sites.

These earth energies help create the "sacred" in "Sacred Ecology." In the next chapter, we will explain the importance of spiritual beings and how they can guide and help you manifest your sacred ecology.

Questions for Group Discussions

- How do you interpret the idea that Earth energies such as ley lines and underground water can alter and uplift human consciousness?
- Have you ever experienced a place with a special energy or presence? How did it affect you? How would you compare your experience to those described in the chapter?
- The concept of "depositing" and "withdrawing" spiritual energy at ley line power centers is intriguing. How do you interpret this idea?
- What are your thoughts on dowsing as a technique to detect Earth energies? Do you find it a credible practice, and why or why not?
- The chapter suggests that humans have an innate sensitivity to detect these energies, which modern technology cannot yet measure. How do you view the relationship between intuition and scientific measurement?
- The chapter highlights the intentional placement of ancient monuments over water lines and ley lines. Why do you think ancient civilizations might have chosen these locations for their monuments?

Chapter 3:
The Power of the Spiritual Beings

Most religions and spiritual traditions recognize and communicate with different types of spiritual beings, such as God, Jesus, angels, saints, Buddhas, deities, elder brethren, and so on. Here is a brief overview of how these beings are recognized and communicated with in different religions:

1. Angels

- Christianity: Angels are messengers of God and serve various roles, including protection, guidance, and delivering God's messages. Communication with angels can occur through prayer, visions, or dreams.
- Islam: Angels are considered messengers and servants of Allah, created from light. They do not have free will and obey Allah's commands. Communication with angels in Islam often involves receiving divine revelations, as in the case of the Prophet Muhammad.
- Judaism: Angels in Judaism are messengers of God, often involved in delivering messages or performing tasks. While direct communication with angels is less emphasized, their presence is acknowledged in various religious texts.

2. Saints

- Christianity (especially Catholicism and Eastern Orthodoxy): Saints are individuals recognized for their holiness and virtuous lives. Catholics and Orthodox Christians may pray to saints to intercede with God on their behalf.
- Hinduism: Although not called "saints" in the same sense as in Christianity, Hinduism reveres holy men and women (such as sadhus, gurus, and rishis) who are believed to have attained a high level of spiritual realization and can intercede for others.

3. Buddhas

- Buddhism: Buddhas are individuals who have attained enlightenment. In Mahayana Buddhism, it is common to venerate Buddhas and Bodhisattvas (enlightened beings who remain in the cycle of rebirth to help others). Devotees may meditate on the qualities of Buddhas, chant their mantras, or make offerings to statues or images.
- Hinduism: Some Hindu sects recognize the Buddha as an incarnation (avatar) of the god Vishnu and may include him in their devotional practices.

4. Deities and Gods

- Buddhism recognizes deities such as Buddhas and Bodhisattvas, e.g., Amitabha, Medicine Buddha, Avalokiteshvara, Manjushri, Tara, Vajrayogini, and others.
- Hinduism: Recognizes a vast pantheon of deities, each representing different aspects of the divine. Worship practices include prayer, rituals, offerings, and meditation. Major deities include Vishnu, Shiva, and Devi (the Goddess), among many others.
- Shinto: The traditional religion of Japan involves the worship of kami, which are spirits associated with natural phenomena, ancestors, and sacred places. Communication with kami is conducted through rituals, offerings, and prayers.
- Ancient Greek and Roman Religions: These ancient polytheistic religions involved worshiping a pantheon of gods and goddesses. Rituals, sacrifices, and prayers were common means of communication with these deities.
- Indigenous Religions: Various indigenous religions around the world have their own pantheons of gods, spirits, and ancestors. Communication with these beings often involves rituals, ceremonies, and offerings.

5. Other Spiritual Beings

- Many New Age beliefs incorporate angels, spirit guides, and other non-physical beings. Practices such as channeling, meditation, and energy work are common ways to communicate with these beings.

Summary

- Angels: Recognized in Christianity, Islam, and Judaism.

- Saints: Venerated in Christianity (especially Catholicism and Eastern Orthodoxy) and analogous figures in Hinduism.
- Buddhas and Bodhisattvas: Central to Buddhism; recognized in certain Hindu traditions.
- Deities and Gods: Worshiped in Buddhism, Hinduism, Shinto, ancient Greek and Roman religions, and various indigenous religions.
- Other Spiritual Beings: Acknowledged in New Age movements and other spiritual practices.

Each of these religions and spiritual traditions has its own unique practices and beliefs regarding the nature and communication with these spiritual beings.

The Spiritual Hierarchy

The Spiritual Hierarchy comprises a great universe and evolution of divine and spiritual beings vital to working with Earth energies and creating places of power and spiritual intention.

All spiritual beings, regardless of religion, want to help us with our spiritual development. I encourage you to learn about the spiritual beings of your spiritual path, faith, or religion.

Communicating and getting guidance from spiritual beings is vital to working with Earth energies and creating places of power, whether a stone circle or a cathedral. The guidance and blessing of the spiritual hierarchy helps us do the right thing at the right place at the right time.

I have developed and built a spiritual relationship with three Spiritual Hierarchies.

1. Devas and the spiritual beings of Nature
2. Buddhas, Bodhisattvas and Buddhist Deities
3. Creators, Grandmothers, and Grandfathers of Native American culture

The word "Deva" means "shining one." Devas are spiritual, nonphysical, beings of light, the architects and engineers of nature, and agents of cosmic law and manifestation.

The Devas

I first became familiar with Devas and the spiritual beings of Nature at the Foundation of Light in Ithaca, New York in the 1970s. At the time, the

Foundation of Light was a spiritual center for meditation and healing and invited healers and spiritual teachers to come, give lectures, and offer workshops. One such teacher was Dorothy MacLean.

Dorothy Maclean was one of the three founders of the Findhorn Foundation, a spiritual community in northeast Scotland. She was best known for her contacts with the consciousnesses of nature she called Devas, embodiments of creative intelligence. Guidance from these beings and their effect on the Findhorn garden was a major element in the fame and growth of the Findhorn community. To learn more about Dorothy and her work with Devas, I recommend the book, *To Hear the Angels Sing: An Odyssey of Co-Creation with the Devic Kingdoms (https://a.co/d/hN0cYD5)*.

In the mid-1970s, Dorothy gave a workshop at the Foundation of Light on how to communicate with Devas. I attended that workshop and was moved and inspired to connect with Devas. I communicate with Devas and the Landscape Angel regularly at Earth Sanctuary. When I dowsed the Seattle ley line map, I began my dowsing session by connecting with the Seattle Deva who guided and protected me.

The Landscape Angel and Spiritual Hierarchy Can Manifest Power Centers

The first ley-line power center at Earth Sanctuary was where the Dolmen is located today. I found the power center by dowsing the Earth Sanctuary property. Not long after that, I visited England and the Avebury Stone Circle. Nearby is the West Kennett Long Barrow, one of the largest, most impressive, and most accessible Neolithic chambered mounds in Britain. As I was walking up the hill to the West Kennett Long Barrow, the Landscape Angel contacted me and said, "Don't worry about finding other power centers at Earth Sanctuary. We will provide the ley line power centers and ley lines; your job is to create sacred spaces."

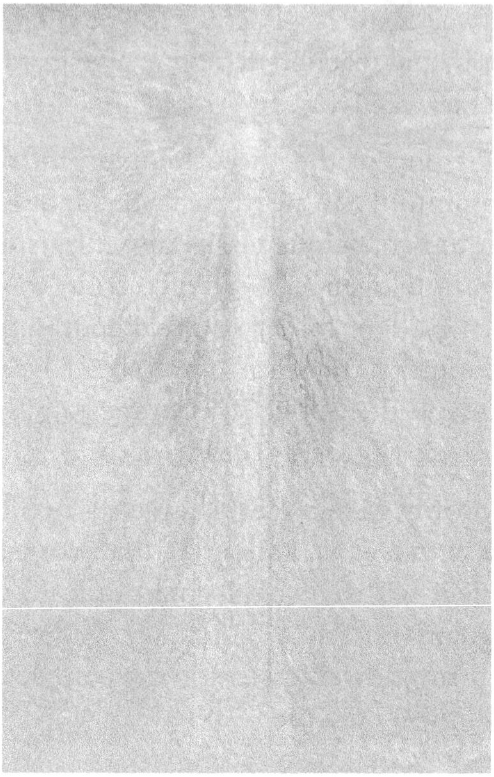

Figure 40 - The Landscape Angel (c)Claudia Pettis

Today, thanks to the Landscape Angel, Earth Sanctuary's sacred spaces are located over ley-line power centers, and ley lines connect and network all the sacred spaces.

When you visit each of the sacred spaces at Earth Sanctuary, close your eyes and feel your sensations; each sacred space seems to have a different feeling based on the experiences of those who have been there before you.

When I visit Earth Sanctuary, I regularly connect with the Landscape Angel to ask her advice and to thank her for her blessings. In my mind's eye, I visualize the Earth Sanctuary Landscape Angel as a white, luminous, glimmering spiritual being fifteen to twenty feet tall, manifested in front of me about six to eight feet off the ground. Perhaps you can feel her presence as well when you visit Earth Sanctuary.

How to Contact and Communicate with Devas and The Landscape Angel

Contacting and communicating with Devas is a holy act and one deserving of respect. It is best to learn deva communication from a person who has experienced this realm. Unfortunately, at the current time, there are few teachers and few books with practical instructions. With that in mind, I believe it may be beneficial to outline my spiritual practice for communicating with devas:

1. **Meditate, clear your mind, and ideate that you are doing this practice for the benefit of all beings.**

Deva contact is a meditation in which one concentrates on the deva with whom communication is desired. Since devas are on a different spiritual level than we are, we must first attune to our inner (divine) self to reach the required vibration level. Use meditation as a vehicle to go into your higher self and to become one with your wholeness.

Visualize and invoke the blessings and help of the landscape angel, your spiritual teachers, and the spiritual beings of your spiritual path or religion, e.g., Chenrezi (the Bodhisattva of compassion).

The most important guidelines here are to follow your inner soul's guidance, be sensitive to your conscience, and discriminate between right and wrong.

2. **Say a prayer of purification, cleansing, and protection.**

You can use one of these prayers of protection or makeup one of your own. You might start by saying:

May I, and all the people near me who might be affected by this activity, be protected from any harm or consequences of this action.

Alternatively, you might like to use the following, which is based on a traditional Christian prayer of protection:

In the Name, through the Power, and by the Word of the Supreme Consciousness/the living Christ/the Buddhas, build a wall of Light and Protection around me, my loved ones, my home, my affairs, and any mode of transportation I may use. I give thanks for this protection. Amen.

Next, I focus my concentration. As I breathe in, I silently say a mantra, or a prayer and I focus all of my concentration on a brilliant point of light—pure spiritual consciousness. Then, as I breathe out, silently saying my mantra, I expand that point of light outward, forming a sphere of spiritual light. I continue this breathing process until a dome of light surrounds and protects the space where I am working.

As you work through this process, you may want to visualize deities of protection all around you.

Please be aware that if it doesn't feel right for you, for any reason, to contact a deva in a specific situation or even to contact devas in general, don't do it.

3. **Ask (using a pendulum): "Can I make deva contact?"**

This is the first of two "checkpoint" questions. Are you qualified and are you able to contact Devas at this time? If the indication is no, respect this answer; it is for your own good and protection. This is no time or place to prove anything.

4. **Ask (using a pendulum): "May I make deva contact?"**

Is it appropriate and right for you to contact Devas, at this time? If the indication is no, do not proceed.

5. **Close your eyes. Think of and visualize the deva you wish to contact. When you think of devas, they are with you.**

Imagine what the deva would look like if you could see it as a being. For example, to contact a flower deva, think of the different colors, shapes, images, smells, and feelings that you associate with that flower. Your focused concentration is like a spiritual lighthouse to the Deva you are seeking.

I have been asked the question, "How do you know who you're communicating with?" One of the reasons to systematically follow this process is to be very safe about who you are communicating with. This is not some trivial psychic game. This is a serious spiritual practice. If you are a meditator and have diligently followed the above steps, you should be communicating with the deva you seek. You can confirm this with your pendulum. For example, you may ask your highest self (through the pendulum), "Is this the pine tree deva?"

A deva can communicate and be sensed in several ways: as a spiritual presence, as a mental image, or even as a smell. In almost all cases, the communication is telepathic and is manifested as messages, thoughts, feelings, or images.

In all cases, keep your thoughts as pure and spiritual as possible to perfectly reflect the light of the angelic being.

You can flow with the mental images you receive in the deva's presence. Or you can take a more active role, by concentrating on specific questions and tuning into the answers that come into your mind. Feel free to make quick notes occasionally so you can refer to them later.

This is a fluid world of archetypes and consciousness. By the power of your concentration, you have reached a level of oneness that transcends the separateness of our bodies. You are now in the last frontier, the frontier of consciousness, an inner universe much vaster than our physical universe. It is a very personal place because you are all alone—a unit of consciousness in a deep sea of cosmic consciousness. Go slowly and carefully. Be responsible. Take care.

6. **Treat the deva with respect, love, and cordiality.**

Devas are divine beings. They respond according to your vibration. Love is the keynote to which the entire spiritual hierarchy responds most strongly. When you feel the presence of the deva near to you, send out waves of love and kindness to it. Imagine the deva is as dear to you as your child, spouse, or parent, and show your love and respect for it through your thoughts and heart feelings.

Those of us who are making some of the first deva contacts in modern times should see deva communication as a special responsibility. We should act as ambassadors for all those who will follow and act as though we are guests in a special place, always on our best behavior. By honoring the divinity of the deva, we are recognizing divinity in the physical flower or another manifest thing. We are seeing the world as beautiful, divine, and wonderful.

7. **End by giving thanks.**

It is good practice in all spiritual practices to end each meditation with a mental sign of thanks or offering. Offer the merits of your actions as a service and benefit for all sentient beings.

In Summary

This process is a very powerful spiritual practice. I recommend adapting it, as summarized below, and using it whenever you dowse as well.

1. Meditate, clear your mind, and ideate that you are doing this practice for the benefit of all beings.
2. Say a prayer of cleansing, purification, and protection.
3. Ask, “Can I do this?”
4. Ask, “May I do this?”

5. Visualize the deva you wish to contact and listen/communicate with an attitude of loving-kindness.
6. Treat the deva with respect, love, and cordiality.
7. End by giving thanks.

What It Feels Like to Communicate with Devas

Here are some experiences that my students have had with devas:

- "A feeling of connection . . ."
- "Devas of old forest groves are those that I feel an affinity with."
- "Saw deva as streaming golden light."
- "When I am around trees, I notice them. I had expectations. As I lose them, I find deva experience."
- "Uplifting communication with tree deva."

Some Helpful Hints

Deva communication is an inner experience, so it is different for each of us. Because it is experiential, it is difficult to precisely explain how to communicate with devas and what happens. It is not a vicarious experience. Devic contact is a very holy act. Don't ever lose sight of this. You may find it helpful to use one deva, who you can contact easily, as an intermediary to contact other devas. For example, I tune into the Landscape Angel to ask general questions or get guidance on how to proceed in a particular area.

When we attune to a deva, our soul begins vibrating at the same level as the deva. By invoking devas, we are calling on those qualities within ourselves. Through interaction with these divine beings, we enlarge our understanding of ourselves and our inner divinity. Use this spiritual technique wisely, for service to all.

Figure 41 - Chenrezi, the Deity of loving-kindness and compassion.

Buddhas, Bodhisattvas, and Deities

My relationship with Buddhas, Bodhisattvas, and Deities is central to Earth Sanctuary and my life. I have been meditating since 1970. My spiritual path is Tibetan Buddhism. For me, it is essential to spend at least one hour every day meditating and doing spiritual practices. My spiritual practices include calm abiding meditations and meditations on Buddhas and Bodhisattvas. For example, one meditation is on *Chenrezi,* the Bodhisattva or Deity of loving-kindness and compassion. In this meditation, I visualize the body of *Chenrezi,* recite his mantra (Om Mani Padme Hung), and imagine Earth Sanctuary and all beings being blessed by *Chenrezi* with love and kindness.

I generally walk the trails of Earth Sanctuary daily. As I walk, I do my spiritual practices, visualizing that all the Buddhas and Bodhisattvas are actually at Earth Sanctuary blessing all the beings that live there as well as visitors and retreatants in the hope that all beings and visitors will quickly become enlightened.

Creator, Grandmothers, and Grandfathers of Native American Culture

The third spiritual hierarchy I would like to mention is the Creator, Grandmothers, and Grandfathers of Native American culture. Native American leaders brought their spiritual culture to Earth Sanctuary. I feel very blessed to have two Native American Medicine Wheels at Earth Sanctuary.

The Medicine Wheel, one of the most powerful and cherished symbols of Native American belief, is an ancient place of prayer sacred to Native Americans. It symbolizes the totality of existence. The Medicine Wheel at Earth Sanctuary is a place for prayer and contemplation, like a portal. It is a sacred space designed to amplify the power of prayers and connect you with all the Grandmothers, Grandfathers, Mother Earth, the Moon, the Sun, the Sky, and the Creator, in the eternal circle of life.

The Medicine Wheel was created by Klaw-osht, an elder and shaman of the Nuu-chah-nulth tribe of Vancouver Island. Prayer flags holding tobacco around the Medicine Wheel are renewed annually with a Native American blessing and consecration ceremony.

During the 2022 blessing ceremony, as I sat outside the Medicine Wheel, I could feel hundreds of grandfathers, grandmothers, and spiritual beings coming to witness and add their power and prayers to the ceremony. This is a blessed and powerful sacred space! When you visit the Medicine Wheel, know that Grandmothers, Grandfathers, and the Creator are there with you, blessing and adding their power and wisdom to your prayers.

Later in 2022, I met a woman who told me about her experience that day in the Medicine Wheel. As she was doing her prayers around the four-colored wheel in the center of the space, she became aware of her own grandmother and grandfather. When she looked around, she told me that all the trees became grandmothers and grandfathers. What a divine experience!

In this chapter, I have discussed the importance of Spiritual Beings and their role in helping create sacred spaces and a sacred ecology. Regardless of your spiritual path or religion, I encourage you to develop and grow a relationship with spiritual beings. This can start just by feeling the blessings of the spiritual beings of your culture and sending them love and appreciation. This is easily done in a garden of flowers, vegetables, and fruits.

My hope is that all visitors of Earth Sanctuary will experience the blessings of the Spiritual Hierarchy at Earth Sanctuary and, most importantly, benefit spiritually from their visit, and share their blessing with others.

~~~

Angels and Trees

Amidst a forest of towering trees,
Lay a pathway of stones with ancient keys.
A trail so serene and winding so well,
Its beauty and peace could surely dispel.

The trees stood tall, like guardians of light,
Shading the stones in coolness and might.
Their leaves rustled in the gentle breeze,
Soothing the heart and bringing peace.

And in the midst of this tranquil scene,
Were the angels, divine and serene,
Their wings like the arms of the trees,
Waving softly in the gentle breeze.

The angels watched over the stones and trees,
Their love and light warm sun on the leaves.
They blessed the land with their heavenly grace,
Guiding the path and lighting up each space.

So if you ever find yourself in this place,
Amidst the trees, the stones, and the angels' embrace,
Take a moment to breathe in the calm,
And feel the love of the angels' balm.

Questions for Group Discussion

These questions can help facilitate a rich and meaningful discussion, encouraging participants to reflect on their own beliefs and experiences while engaging with the themes of the chapter.

- What are your thoughts on the concept of communicating with spiritual beings?
- Have you ever felt a spiritual connection to nature? If so, can you share your experience?
- The author incorporates elements from Tibetan Buddhism, Native American traditions, and other spiritual practices. How do you feel
~~~

about blending different spiritual traditions? Can this integration enrich one's spiritual experience, or does it risk diluting individual traditions?

- How do personal spiritual experiences shape your understanding of the divine or the spiritual world?
- Have you tried any form of meditation or visualization? What effects did it have on you?
- How can the idea of seeking guidance from spiritual beings be applied in everyday life?

Chapter 4: The Power of Ceremony to Bless and Heal

Throughout history, ceremonies have played a significant role in our cultural and spiritual practices. These rituals are not mere formalities but powerful tools for blessing and healing. Whether it's a prayer, a seasonable festival (e.g., solstice), a memorial ceremony for a loved one, or a way to honor a special occasion with family and friends, ceremonies hold the potential to transform lives and foster a deeper connection to ourselves, others, and the universe. Ceremonies can be spiritual, religious, cultural, or social and often involve specific rituals, symbols, or actions that carry deep meaning and significance.

Ceremonies play a crucial role in the creation of sacred ecologies, and sacred spaces, including temples and churches, for many reasons:

- **Create Sacred Space**: Ceremonies help to transform an ordinary physical location into a sacred or spiritual place. Through rites, prayers, and blessings, the space is imbued with spiritual significance, setting it apart from the mundane world.
- **Consecrated Space**: Many spiritual and religious traditions require specific rituals to formally consecrate or dedicate a space to a divine purpose. For example, the consecration of a church may involve the anointing of the altar, the blessing of the building, or the installation of religious symbols. This sanctifies the space for worship and community gathering.
- **Foster Communal Experience**: Ceremonies are communal events that bring people together in a shared act of faith and devotion. Involving the community in the process of creating or blessing a sacred space, fosters a sense of collective identity and shared responsibility for maintaining the spiritual significance of the place.

- **Affirm Symbolism and Tradition**: Spiritual and religious ceremonies are rich in symbolism, often drawing on centuries of tradition. The acts performed during these rituals (e.g., lighting candles or reciting sacred texts) carry deep meaning and reinforce the spiritual atmosphere of the space. This continuity of tradition links the present with the past, affirming the sacredness of the space.
- **Mark Transitions and Milestones**: Ceremonies often mark important life events such as births, marriages, graduations, and funerals. These events symbolize transitions from one stage of life to another, providing a structured way to acknowledge and celebrate these changes. Earth Sanctuary's Memorial Tree program helps people honor a loved one and also celebrate significant life events.
- **Invoke the Divine Presence**: Ceremonies often involve prayers, invocations, or chants to invite the presence of the divine into the space. This act of inviting a higher power to dwell within the sacred space reaffirms its purpose as a spiritual center, fostering a sense of connection between the divine and the human community.
- **Preserve Cultural, Spiritual, and Religious Identity**: Ceremonies play a vital role in preserving and transmitting cultural, spiritual, and religious traditions. They help individuals and communities maintain a connection to their heritage, passing down values, beliefs, and practices from one generation to the next.
- **Strengthen Community and Social Bonds**: Ceremonies bring people together, fostering a sense of community and belonging. Whether it's a wedding, a holiday celebration, or a memorial service, these events create shared experiences that strengthen social ties and reinforce group identity.
- **Provide Emotional and Psychological Support**: Ceremonies provide a structured way to express and process emotions, especially during times of joy or sorrow. They offer comfort, closure, and a sense of order during significant life events, helping individuals and communities cope with change and uncertainty.

Overall, ceremonies are important because they give meaning to significant life events, strengthen social and cultural bonds, and help individuals and communities navigate the transitions and challenges of life. I encourage readers to bring more ceremony into your life.

How to Plan a Ceremony

- Work with a small group of people and get them involved at the beginning. Involvement is the key to a successful ceremony. Explain the parts of the ceremony and give each person an activity such as reading a poem, smudging, lighting candles, or placing flowers. Ask all to have an open mind, as this kind of ceremony may be new to them.
- Pick a special calendar time of the year. Research the history, background, and tradition of that day.
- Choose a single leader to integrate everybody's ideas into a cohesive program. This person should write up and distribute copies of the meaning and symbolism of the day as well as a simple description of the ceremony itself. Hand this program out as people arrive so they can read and assimilate it.
- The real power of the ceremony is what people are thinking and feeling during the ceremony. Use offerings as symbolic gifts to the higher self and the deities. These include candles, incense, semiprecious stones (like quartz, obsidian, or lapis), bones, flowers, fruit, and so on.
- Involve the kids. They'll love it and their happiness is contagious!
- If possible, include live music. Poetry is also extraordinarily effective.
- Have food and drink afterward.
- Make it a yearly tradition—do it every year at about the same time.

Ceremony at the Mutiny Bay Stone Circle

In 1995, I designed and built the Mutiny Bay Stone Circle using the same design elements and techniques used in most of the world's ancient monuments (e.g., Stonehenge, Avebury, and the Great Pyramid). The Mutiny Bay Stone Circle is located over a ley line power center and an underground blind spring. Two ley lines came in while the stone circle was in construction.

Figure 42- The Mutiny Bay Stone Circle in 1995 on Whidbey Island. (c) Chuck Pettis

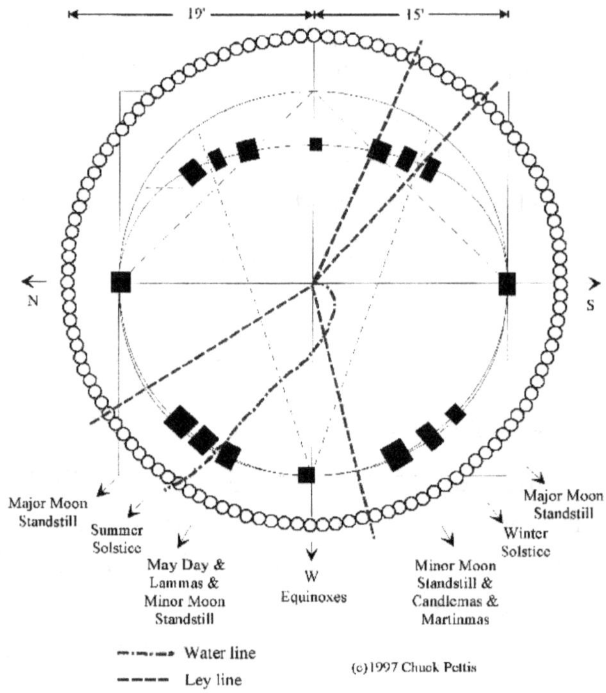

Figure 43 - I designed the Mutiny Bay Stone Circle based on a squared-circle geometry such that the perimeter of the square that circumscribes the inner stone circle is the same length as the circumference of the outer circle: 120 feet. The Cottonwood Stone Circle at Earth Sanctuary also has a circumference of 120 feet. (c) Chuck Pettis

Figure 44- Sunset on May Day. (c) Chuck Pettis

The Mutiny Bay Stone Circle has something that isn't present at Earth Sanctuary: a horizon. Every stone marks an important astronomical alignment: Summer Solstice, May Day, Lammas, Equinoxes, Candlemas, Martinmas, Winter Solstice, and also the Major and Minor Moon standstills. Each of these alignments has a holiday associated with it and a ceremony.

When I was designing the Mutiny Bay Stone Circle, I was guided by the Landscape Angel, the Bodhisattva Chenrezi, Supreme Consciousness, Shrii Shrii Anandamurti, the Earth Spirit, the Ancient Ones (designers and builders of ancient monuments), and devas of Icy, water, stone, sand, wind, ocean, and evergreen trees. Every stone was asked if belonged in the stone circle. I visualized an ascending pyramid of Buddhas and Bodhisattvas directly above the stone circle.

H.H. Jigdal Dagchen Sakya led a blessing and consecration ceremony at the Mutiny Bay Stone Circle. After the ceremony, H.H. Jigdal Dagchen Sakya told me that many insects were killed while I was building the stone circle, so he did a fire puja (ceremony) and mantras for the benefit of all the insects. I had never thought of this before and have become much more aware of how all the little insects and beings are harmed by my work. For example, I now recite the Aspiration of Samantabhadra meditation text for the sake of all the beings that might be harmed when we disturb the earth or when we bury ashes by a Memorial Tree at Earth Sanctuary.

Figure 45 – A white scarf (kata) was placed on the large standing stone that marks the Equinoxes by His Holiness Jigdal Dagchen Sakya. (c) Chuck Pettis

Figure 46 - After the consecration ceremony, the blessed rice was placed in the middle of the stone circle. (c) Chuck Pettis

Deity Yoga Ceremony

In a ceremony, we participate in the acting out of deep, intuitive visualizations relating ourselves to something greater than ourselves that facilitate the gradual assimilation of spiritual values into one's life. In Tibetan Buddhism, we do this by practicing Deity Yoga, identifying ourselves as an archetypal Buddha.

At Sakya Monastery, the main Deity Yoga practice is centered on Chenrezi, the embodiment of perfect love, kindness, and compassion. Practitioners visualize Chenrezi as a radiant, smiling Buddha and then identify with him. Through this visualization, they seek blessings for purification, meditative concentration, and enlightenment, reciting the mantra "Om Mani Padme Hung" and engaging in calm abiding meditation. The practice concludes with a dedication prayer, wishing for all beings to attain Buddhahood, emphasizing altruism and the karmic power of working for others' happiness as the foundation of future joy.

Examples of Blessings Ceremonies at Earth Sanctuary

The lamas of Sakya Monastery in Seattle have done many blessings and ceremonies at Earth Sanctuary. His Holiness Jigdal Dagchen Sakya, the head lama of Sakya Monastery of Tibetan Buddhism (www.sakya.org) gave teachings on *Ngondro,* Medicine Buddha, performed a fire puja ("act of worship") ceremony in the Cottonwood Stone Circle and led the seven ceremonies at the Earth Sanctuary Stupa.

Figure 47 - One of seven Tibetan Buddhist blessing ceremonies held at the Earth Sanctuary Stupa. (c) Chuck Pettis

One of the ceremonies done during the construction of the Stupa was a Vajrapani fire puja. Vajrapani is a Tibetan Buddhist guardian deity symbolizing the Buddha's vitality, power, and fearlessness. A fire puja is an elaborate or simple ceremony in which the Lama (spiritual teacher) invokes a deity and makes offerings to honor the deity and receive the blessings of the deity.

Figure 48 - Tibetan monks created a sand mandala on this steel plate. (c) Chuck Pettis

Figure 49- A sacred ceremonial fire was lit on top of the sand mandala. (c) Chuck Pettis

Some said they could see Vajrapani in the fire. Afterward, the steel plate was placed in a shed at Earth Sanctuary. But I kept getting inner messages that the steel plate wanted to be utilized. So, my wife, Claudia, and I created the Veils of Reality sculpture that incorporates the painted steel plate. Some people have reported seeing unusual and magical images of themself while standing in front of it.

Figure 50- The consecrated steel plate was resurrected into a new art form, uniting the magic of Tibetan Buddhism with modern minimalistic art. (c) Chuck Pettis

Stand in front of the left black steel plate. What do you see? We have observed that everyone sees something different and something special. Ultimately, if we look hard and long at a pure piece of art, we will be looking at ourselves in a new way for the first time. Enjoy!

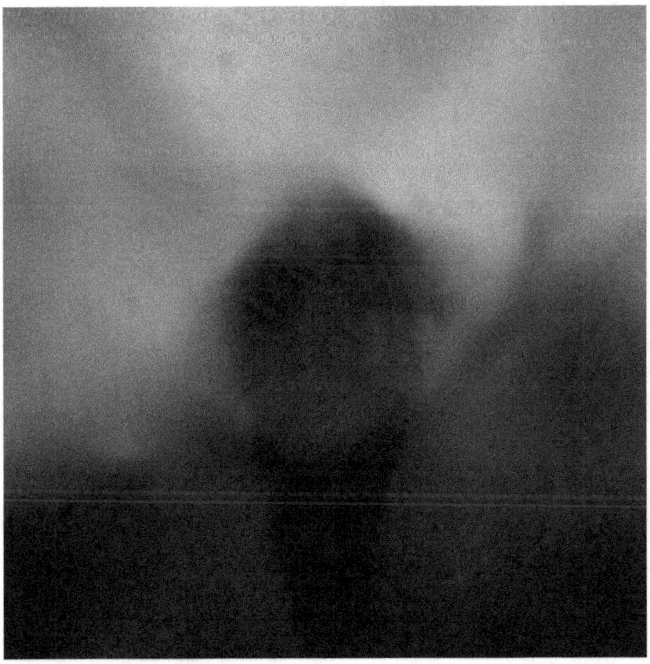

Figure 51 - Reflection of a person in front of the steel plate on the left side of the Veils of Reality sculpture. (c) Chuck Pettis

Figure 52 - Ceremony by H.E. Dagmo Kusho Sakya at the Fen Stone Circle. (c) Chuck Pettis

H.E. Dagmo Kusho Sakya, wife of H. H. Jigdal Dagchen Sakya and a leading female lama, gave Tara (an important female Buddha) teachings in the Fen Stone Circle and at the Dolmen. H.E. Avikrita Sakya, H.E. Abhaya Sakya, H.E. Asanga Sakya, the next generation of Sakya lamas, leaders and teachers, and the monks of Sakya Monastery have all visited and blessed Earth Sanctuary.

Rev. Lawrence Koichi Barrish, Honorable High Priest of Tsubaki Grand Shrine of America (https://tsubakishrine.org/) conducted a blessing ceremony in the Cottonwood Stone Circle, sprinkling sake on all the stones.

Figure 53- H.E. Dza Kilung Rinpoche with Chuck Pettis (c) Chuck Pettis [Photo taken by family of Kilung Rinpoche]

Figure 54- Monks from Atammayataram Monastery blessing ceremony. (c) Chuck Pettis

Figure 55- Klaw-osht, an elder and shaman of the Nuu-cha-nulth tribe of Vancouver Island has done ceremony in the Cottonwood Stone Circle for the Medicine Wheel and also did a ceremony at the Baby Gray Whale Medicine Wheel. (c) Chuck Pettis

Figure 56- Celia Sullivan, Grandmother Vi Hilbert, and Norma Jean Young at the Dolmen. Photographer unknown.

Grandmother Vi Hilbert, a revered Upper Skagit/Salish Grandmother, teacher, and keeper of Native American wisdom, speaking in her native language, offered words of gratitude to the spirits of the megalithic stones of Earth Sanctuary's dolmen. Thanking them for their strength and serenity, she added "You will be here long after we have been gone. You will be here to do work for generation after generation."

Figure 57- Father Bill Johnson, a Jesuit priest, celebrated a version of Teilhard de Chardin's "Mass of the World" in the Cottonwood Stone Circle. (c) Chuck Pettis

Father Bill Johnson, a Jesuit priest, celebrated a version of Teilhard de Chardin's "Mass of the World" in the Cottonwood Stone Circle. Pierre Teilhard de Chardin, a French philosopher and paleontologist, is known for his theory that man is evolving, mentally and socially, toward a final spiritual unity.

Many other spiritual and religious leaders and practitioners have held ceremonies during a retreat or on their own. Earth Sanctuary welcomes blessing ceremonies by other religious and spiritual leaders.

When a spiritual leader performs a ceremony, the blessings remain in the land for the benefit of both human visitors and all the inhabitants of Earth Sanctuary. The power of ceremony to bless and heal is rooted in its ability to provide structure, meaning, and connection in our lives. Whether it's for healing from loss, celebrating life's milestones, connecting with our cultural or spiritual heritage, or invoking blessings, ceremonies are a powerful tool that transcend time and culture. They remind us that our lives are filled with sacred moments and that we can find healing and blessings in the most ordinary and extraordinary of events. Embracing the power of ceremony can enrich our lives and help us navigate the complex journey of being human.

At Earth Sanctuary, we encourage visitors and retreatants to conduct ceremonies and celebrations of their own spiritual or religious tradition, not just for themselves, but for the benefit of others.

The Power of Making Offerings

I have an altar in my meditation room with symbols of enlightenment, including statues of Buddha, buddhas, deities, bodhisattvas, wealth vases, and stupas. I make offerings at the altar daily using water bowls representing water for drinking, water for washing the feet, flowers, incense, perfumed water, and food. As I fill the water bowls, I recite Om, Ah, Hum imagining that the offerings are received by all the holy objects on my altar. All the offerings are made to help all sentient beings attain enlightenment as soon as possible.

An altar is beneficial in many spiritual, religious, and personal contexts because it serves as a focal point for worship, reflection, and connection to the divine. It acts as a sacred space where one can offer prayers, meditate, and honor spiritual practices. Here's why an altar is beneficial and how someone can create one:

Benefits of an Altar

- **Spiritual Focus**: An altar creates a dedicated space for spiritual activities such as prayer, meditation, or rituals. It helps individuals focus their attention, channel their energy, and cultivate a sense of inner peace.
- **Symbol of Devotion**: An altar serves as a physical manifestation of one's faith or spiritual commitment. It can be a place to express gratitude, offer thanks, and acknowledge the presence of the divine in everyday life.
- **Connection to the Divine**: In many traditions, an altar is viewed as a point of connection between the human and the divine. It can help facilitate communication with higher powers, ancestors, or spiritual beings.
- **Center for Ritual and Worship**: An altar is central to many religious or spiritual rituals. It often contains symbols, images, or objects that are meaningful to a particular practice, enhancing the ritual's significance.
- **Personal Empowerment and Healing**: By creating and maintaining an altar, people often feel more empowered, grounded, and connected to their inner selves. It can also be a space for healing, where one can let go of negative energy and invite positive, nurturing forces.
- **Cultural and Ancestral Connection:** For many, an altar is a way to honor ancestors, cultural traditions, or spiritual heritage. It can serve as a reminder of one's roots and a way to seek guidance or strength from past generations.

How to Create an Altar

- **Choose a Space**: Select a quiet, personal spot in your home or outdoor environment where you can create a sacred space. It should be a place where you feel peaceful and where you can easily access it for daily rituals or meditation. A small table, shelf, or even the ground can serve as the base for your altar.
- **Set Your Intention**: Determine the purpose of your altar. It could be for daily meditation, prayer, honoring ancestors, a seasonal altar, or a space for a specific spiritual or religious practice.
- **Add Sacred Objects**: Place objects that hold spiritual and symbolic significance for you—statues, candles, crystals, stones, sacred books, or religious icons. Incorporate elements like flowers, plants, water bowls, or sand to enhance the natural energy of the altar. Personal items like photos of loved ones, mementos, or personal objects can make the altar feel more connected to your life.

- **Make Offerings**: Many people include offerings such as incense, candles, or food to symbolize their devotion or gratitude.
- **Arrange Items Mindfully**: Arrange objects in a way that feels balanced to you. Many people choose to have a central object, like a statue or candle, as the focal point. Some people like to represent the four elements—earth, air, fire, and water—by placing objects like stones (earth), feathers (air), candles (fire), and water (water).
- **Keep it Clean and Sacred**: Regularly clean your altar, refresh offerings, and update it as needed. This keeps the energy flowing and honors the sacredness of the space.
- **Regular Ritual**: Establish a regular practice of visiting your altar, whether it's lighting a candle, offering a prayer, or simply spending time in reflection.
- **Personalize It**: Your altar should reflect your beliefs and needs. There are no strict rules, so feel free to customize it in ways that feel meaningful to you, whether through colors, themes, or objects that resonate with your spiritual journey.

Creating an altar is a deeply personal process, and its power comes from the intention and care you put into it. It can serve as a sanctuary where you reconnect with your spirit, set your intentions, and find peace in daily life.

The Symbolism of Offerings at Earth Sanctuary

While visiting the sacred spaces of Earth Sanctuary, you will see symbolic offerings that people have left such as crystals, handwritten notes, incense, flowers, and more. Offerings are a remedy for our lifetime of "greed" and "attachment" in the same way that bowing, or prostrations are a remedy for "pride" and "ego. Offerings reinforce the activity of giving — and the positive merit giving accumulates. When you "give," you create an opening for "receiving."

Figure 58 - Offerings on the altar at the Earth Sanctuary Dolmen. (c) Chuck Pettis

The Power of Giving: Offerings provide a unique opportunity to generate merit. In Christianity, it is taught that God desires us to give because of the blessings that flow from generous acts toward Him and others. Similarly, in Buddhism, making offerings is a spiritual practice for accumulating positive potential and purifying clinging and miserliness.

The Magic of Right Intention: The transformation in our mind that occurs when we make an offering creates positive potential—meritorious good karma—which becomes the cause for future happiness. This is the profound magic of giving with the right intention.

A Call to Bring Your Offering: When you visit Earth Sanctuary or another sacred place, consider bringing a small offering that holds personal meaning and symbolic significance for you. Place your offering in the sacred space of your choosing as a gesture of connection and gratitude.

Questions for Group Discussions

These questions can help guide a meaningful and insightful discussion about the chapter, allowing participants to explore their own experiences and beliefs in relation to the themes presented.

- Have you ever participated in a ceremony that had a significant impact on you? Can you share your experience and how it affected you?

- Why do you think ceremonies have been a part of human culture throughout history?
- What are some ceremonies you participate in regularly, and what do they mean to you?
- The author mentions receiving guidance from various spiritual beings while designing the stone circle. What are your thoughts on this type of spiritual guidance?
- The author welcomes ceremonies from different spiritual and religious traditions at Earth Sanctuary. How does this inclusivity enrich the sanctuary and its visitors?
- Do you have an altar? If so, what does it mean to you?
- What offerings would you consider bringing to a sacred space, and why?
- How can you incorporate the principles and practices of ceremonies into your daily life to enhance your spiritual well-being?
- What steps can you take to deepen your connection to the natural world through ceremonies or rituals?

Chapter 5:
The Power of the Stupa, Prayer Wheels, and Prayer Flags

"Holy objects like the Earth Sanctuary Stupa have the power to bring immeasurable spiritual good and blessings."

The stupa is the most important Buddhist monument and a sacred space. It represents Buddha's holy body, speech, and mind and the path to becoming a Buddha and enlightened being.

Figure 59 - The Earth Sanctuary Stupa (c) Chuck Pettis

The shape of the stupa represents Buddha, crowned and sitting in a meditation position.

1. The square base is the ethical foundation and his throne.
2. The three circular rings in the middle are his body.
3. His head is the domed section.
4. The 13 gold rings are the levels of spiritual attainment.
5. The umbrella is the great compassion of the Buddha.
6. The moon and sun represent the intention to relieve all beings of suffering and to attain the happiness of enlightenment.
7. The pinnacle at the top is the crown of Buddha.

"Earth Sanctuary is an ideal location to transform your mind to a brighter and more compassionate state. When I engaged in the morning meditation, I would take time to circumambulate around the blessed stupa to say my mantras. I discovered the local wildlife will subtly participate in your experience. As a woodpecker hammered away on the side of a nearby tree, I imagined it was the playful sound of a Tibetan damaru drumming in support."

Figure 60 - The Usnisa Vijaya statue. Usnisa Vijaya is one of the three long-life Buddhist Deities. (c) Chuck Pettis

The Earth Sanctuary Stupa has three chambers, each filled with sacred scriptures, prayers, dried flowers, and other holy objects such as Tsa Tsas which are clay statues of Buddhas and deities. An important element in every stupa is the Tree of Life, a wooden pole inscribed with many prayers and placed vertically in the center of the stupa.

There are eight types of stupas. The Earth Sanctuary stupa is a Victory Stupa that commemorates Buddha's successful prolonging of his life by three months. Circumambulating the Victory Stupa at Earth Sanctuary is said to extend your life.

According to Buddhist teachings, a stupa helps calm the mind, radiates spiritual blessings, amplifies prayers, promotes long life, attracts wealth, helps cure illness and disease, and plants the seeds of enlightenment.

"Circumambulating the Stupa was meaningful and healing, calming, and centering."

Figure 61 - After you spin the prayer whales clockwise, walk around the Stupa clockwise (circumambulate) with prayers for the happiness and relief of suffering of all beings. (c) Michael Stadler

If we have difficulties in life and want to achieve success, we need to purify the negative karma that caused these difficulties and create merit for future success and happiness. Walk around the stupa clockwise saying prayers, mantras, or thinking spiritual thoughts. This method can be used to find solutions to many problems.

Suggested Prayers and Chants to Use When Circumambulating the Stupa

- Pray: "May all beings be happy. May all beings be free of suffering."

- Say a prayer from your own spiritual or religious tradition. The stupa will amplify your prayers.
- Imagine white light is coming from the stupa, totally illuminating and purifying you from sickness and negative karma.
- Chant the most popular Tibetan Buddhist mantra: Om Mani Padme Hung. This is the mantra of Chenrezi, the Boddhisatva of loving-kindness and compassion. Here is a more detailed explanation from His Holiness the Dalai Lama[6]:
    - The first, Om is composed of three letters. A, U, and M. These symbolize the practitioner's impure body, speech, and mind; they also symbolize the pure exalted body, speech, and mind of a Buddha.
    - Mani, meaning jewel, symbolizes the factors of method—the altruistic intention to become enlightened, compassion, and love. Just as a jewel is capable of removing poverty, so the altruistic mind of enlightenment is capable of removing the poverty, or difficulties, of cyclic existence and of solitary peace. Similarly, just as a jewel fulfills the wishes of sentient beings, so the altruistic intention to become enlightened fulfills the wishes of sentient beings.
    - Hum. Purity must be achieved by an indivisible unity of method and wisdom, symbolized by the final syllable hum, which indicates indivisibility. According to the sutra system, this indivisibility of method and wisdom refers to wisdom affected by method and method affected by wisdom.

Benefits of Circumambulating a Stupa, According to Buddhist Tradition

"The stupa was a rejoicing moment of stability and heavenly broadcast of endless bliss."

For those with a disease that is difficult to cure or incurable, such as cancer, it is very beneficial to circumambulate a stupa several hundred times a day. This is said to purify the cause of the disease.

- When rain touches the stupa, it becomes holy water. When the rainwater runs from the stupa and soaks into the ground, any insects, worms, slugs, or any living in the ground get purified and will receive a higher rebirth and become liberated. It is the same with the wind. When the wind blows over a stupa, it becomes blessed and benefits all beings downwind.

- Stupas are more precious and wish-fulfilling than skies filled with gold, diamonds, or even wish-fulfilling jewels because the stupa has a lasting positive effect on your mind. Bottomline, the stupa makes it easier for us to train our minds so that we begin to act like a Buddha or holy person.
- The temporal and ultimate benefits are infinite. Circumambulating a stupa is the supreme method to purify obscurations and close the door to the lower realms.
- The sutra called Circumambulating the Stupa says: By circumambulating stupas, one avoids being born an animal, a barbarian, or a fool. One will have a long life. For hundreds of eons, one will not be born blind or crippled with arthritis. By circumambulating stupas, one receives perfect power and perseverance; because there is no laziness, one can develop the mind quickly. One receives six clairvoyances. Finally, one achieves the golden holy body of Buddha, adorned with holy signs.

Building the Earth Sanctuary Stupa

There is no "cookbook" for building a stupa; it is a complicated undertaking. Preliminary architectural drawings were provided by Jhampa Shaneman. Matthew Swett of Taproot Architects prepared detailed drawings for the Earth Sanctuary Stupa.

When building a stupa, guidance, direction, and ceremonies from a Buddhist teacher are necessary. We are deeply grateful to His Holiness Jigdal Dagchen Sakya, one of the revered Tibetan Buddhist leaders, who oversaw the design and construction of the Earth Sanctuary Stupa. He led the initial "Ground Ceremony," as well as the final "Blessing and Consecration Ceremony." We are also thankful for those assisting H.H. Jigdal Dagchen Sakya: H.E. Dagmo Kusho Sakya, Ven. Tulku Yeshi, Ven. Khenpo Jampa and Lama Migmar.

We would also like to thank Tarthang Rinpoche, Khenpo Tsering and the College of Dialectics, the Sullivan family, Tim Scriven, Steve Haynes, Albert Gabelein, Candy Young, and many other kind and generous people who made the Earth Sanctuary Stupa possible.

Figure 62 - Seven Tibetan Buddhist ceremonies were performed during the construction of the stupa. (c) Chuck Pettis

The Earth Sanctuary Stupa was expensive to build. We welcome donations to help cover the cost of its building. To contribute to creating and building a stupa is of immense benefit. You can dedicate the merit of your contribution to family members who have passed away, to people and beings who are ill or suffering, or to world peace.

Thank you for your meditations, prayers, and blessings at the Earth Sanctuary Stupa.

The Most Powerful Prayer Wheels in the World Are at Earth Sanctuary

In the old days of Tibet, people everywhere, particularly older people, spun prayer wheels from morning to night, while reciting the *Om Mani Padme Hum*[7] mantra (power words) to relieve the misery of all beings and bring them happiness.

Prayer wheels are cylinders filled with many mantras (Tibetan Buddhist prayers) mounted on a spindle and spun clockwise with intention and prayer. The word *mantra* has Sanskrit linguistic roots whose literal meaning is "protection of the mind." In Tibetan Buddhism reciting mantras is one of the most effective ways in which a person can actively create a peaceful, relaxed, and happy state of mind.

The primary goal of the traditional Tibetan Buddhist prayer wheel practice is to increase happiness and relieve the miseries of all beings. As prayer wheels rotate clockwise, Buddhists believe that they radiate blessings for all beings. We are taught that peace, happiness, kindness, and relief from suffering for all beings are radiated through the blessings of the mantras inside the prayer wheel, and by the intention and concentration of the spiritual practitioner.

Buddhist texts teach that mindfully turning a prayer wheel produces the same merit and benefits as having recited the number of prayers (mantras) inside the prayer wheel multiplied by how many times the prayer wheel spins around.

The Earth Sanctuary Tibet Tech Prayer Wheel

Figure 63- This Tibet Tech Prayer Wheel is located by the Cottonwood Stone Circle.

Prayer wheels have traditionally been filled with mantras carefully printed on paper or microfilm and then wound up inside. The largest prayer wheel in Old Tibet contained 100 million prayers printed on paper with hand-carved wood blocks.

A High-Tech Prayer Wheel

The idea for a "high-tech" prayer wheel came from a meeting I had with H.H. Jigdal Dagchen Sakya and his wife, H.E. Dagmo Kusho Sakya. They recommended that I use prayer wheels in my daily meditation practice and gave me two prayer wheels as gifts. One was electric and revolved continuously on my home shrine. The other prayer wheel was held in my hand and spun while I did my spiritual practices. When I started spinning the prayer wheels, my wife immediately noticed an increase in the spiritual feeling around our altar. One day, while meditating, the question arose in my mind, 'How many prayers (mantras) could be put in a prayer wheel?' In answer, I created a new state-of-the-art spiritual technology, which I branded "Tibet Tech®."

His Holiness Jigdal Dagchen Sakya, one of the most revered Tibetan Buddhist masters, oversaw the design and construction of the prayer wheels and blessed and consecrated the DVDs used in Tibet Tech prayer wheels. DVDs are used to maximize the number of mantras that can be placed within the prayer wheel cylinder. The power of the prayer wheel comes from one's spiritual intention and the blessings of H.H. Jigdal Dagchen Sakya.

At the beginning of the project, H.E. Dagmo Kusho Sakya and Chuck Pettis contacted Head Lama Tarthang Tulku Rinpoche of the Tibetan Nyingma Meditation Center, which has created over 1,600 prayer wheels. Tarthang Tulku was kind enough to provide a very helpful article entitled "Prayer Wheels of the Nyingma Center."

Constructing the Tibet Tech Prayer Wheel

The construction of the unique Tibet Tech spiritual tools builds on traditional prayer wheel design techniques by incorporating modern computer technology. Tibet Tech prayer wheels are the first to use DVDs to store mantras inside a prayer wheel. H.H. Jigdal Dagchen Sakya selected the mantras, Ven. Yeshi Tulku typed the mantras, and H.H. Jigdal Dagchen Sakya and H.E. Dagmo Kusho Sakya proofread the mantras. Each of the eight mantras contained in the Tibet Tech prayer wheels was placed by Tom Ashbrook, a Sakya Monastery member, into an evenly spaced 10 by 5000 cell Microsoft Excel worksheet matrix, which was printed to an Adobe Acrobat PDF file containing 50,000 mantras. One thousand of these PDF files were compressed into a single zip file containing 50 million mantras. For each of the eight mantras, as many of the zipped files as possible were copied to a DVD. The DVDs do not play like a movie; the DVD is used to maximize the storage capacity of the prayer wheel.

The DVD duplicator uses lasers to read and write mantras onto an opaque reflective layer in 0.4-micron pits in a spiral track. "Although that is a technically accurate description of the DVD-RW process," explains Tom Ashbrook, "while producing the DVDs, it looked to me like the mantras were being written by light onto a rainbow onto the DVD."

Below and above the stack of DVDs are the Earth Wheel and Sky Wheel symbols, as drawn by Lama Zopa Rinpoche and depicted in the book, Wheel of Great Compassion: The Practice of the Prayer Wheel in Tibetan Buddhism by Lorne Ladner (Wisdom Publications, 2000).

The prayers on all the DVDs used in each Tibet Tech prayer wheel are blessed and consecrated by H.H. Jigdal Dagchen Sakya in a three-part blessing ceremony. The first part is a praise of the mantras, then the DVDs are consecrated. Finally, a Dedication of Merit prayer is recited to benefit all sentient beings and to work for a more peaceful world. Dagchen Rinpoche said, "As long as the universe exists, these mantras will remain potent."

In 2004, two special custom-built Tibet Tech prayer wheels were created by artist and metal craftsman Mike Sweeney with some engineering assistance from Charlie Pettis (my father) and then installed at Earth Sanctuary, where they have been spun over 4 million times by retreatants and visitors. That's 1.3 trillion x 4.0 million (5,200,000,000,000,000,000) prayers radiated to benefit all living beings and help bring peace to Earth.

The Earth Sanctuary Tibet Tech Prayer Wheels are designed in a traditional Tibetan Buddhist style -- the housing being one that you might see in Nepal, India, or Tibet, and often seen around monasteries and temples.

1.3 Trillion Prayers per Revolution!

Each revolution of the Tibet Tech prayer wheel releases 1.3 trillion prayers, radiating an incalculable amount of merit, peace, kindness, and relief from suffering to all beings. Mindfully turning a prayer wheel with 1.3 trillion mantras produces the same merit and benefits as having recited 1.3 trillion mantras At one mantra recitation per second, it would take a person approximately 42,776 years to recite 1.3 trillion mantras – just one revolution of the prayer wheel!

Each Tibet-Tech prayer wheel includes sixteen sets of eight magical and powerful mantras (128 DVDs in total). The mantras are stacked from the bottom up, starting with the A-li mantra; the Ka-li mantra, and so on up to the eighth mantra, the 100-syllable Vajrasattva mantra.

Prayer (*Mantra*)	Benefits	Prayers per DVD
1.- A-li	The foundation of all mantras.	21,750,000,000
2- Ka-li	Purifies speech.	7,000,000,000
3- *Om Tara Tuttare Ture Soha*	Immediate relief from suffering.	12,650,000,000
4- *Om Mani Padme Hum*	Relieves suffering of all beings	14,400,000,000
5- *Om Amarana Tsibin Tiye Soha*	Long life.	13,400,000,000
6- *Om Ah Hum Bendza Guru Padma Siddhi Hum*	Grants divine powers and realizations.	9,800,000,000
7- Ye-dhr-ma	Consecration and stabilization.	5,250,000,000
8- Vajrasattva	Purification	98,750,000
Totals		84,348,750,000

Figure 64 - What is inside the Tibet Tech Prayer Wheel.

Benefits of Prayer Wheels

The practice of spinning a prayer wheel has many "magical" benefits. According to tradition, the prayer wheel was brought to Earth from the realm of Nagas (dragon-like beings who live in oceans) by the great Bodhisattva Nagarjuna because he was told by Avalokiteshvara (also known as Chenrezi, the Bodhisattva of love, kindness, and compassion) in a vision that "the benefits to living beings will be enormous." Nagarjuna gave the prayer wheel practice to the Lion-Faced Dakini (a female spiritual being), who in turn gave it to Padmasambhava (the second Buddha) who brought it to Tibet.

Sakyamuni Buddha said that turning a prayer wheel once is better than having done years of retreat, and intensive spiritual practice. He said, "Turning the Dharma wheel is better than listening, reflecting and meditating for eons[8]."

The use of the prayer wheel is one of the easiest ways to purify past negative karma, non-virtuous actions, defilements, and obstacles that prevent us from realizing our true self and understanding the true nature of reality. Other prayer wheel benefits include:

1. Transforming one's home and property into a very peaceful, pleasant, holy land or "high heavenly realm."

2. Saving all the beings in the area around prayer wheels from rebirth in the lower realms (e.g., animal incarnation).
3. Purifying our body, speech, and mind.
4. Accumulating extensive merit for oneself and all beings in the area.
5. Preventing harm from spirits and negative beings.
6. Healing sicknesses and protecting people from contagious diseases and epidemics.

Proper Visualizations While Spinning a Prayer Wheel

Spinning a prayer wheel is not a mindless exercise. When spinning the Tibet Tech prayer wheels at Earth Sanctuary, do so with the proper intentions. The prayer wheel practice is visualized as a manifestation of the body, speech, and mind of the Buddha.

With our hands (body) we spin and move the prayer wheel. With our speech, we recite one of the mantras in the prayer wheel (e.g., Om Mani Padme Hum). And with our mind, we engage in visualizations or recitations, using our motivation and intention to bless all beings and bring peace to our surroundings and the entire world.

What should you do while spinning a prayer wheel?

- Spin the prayer wheel clockwise mindfully with the intention and prayer: "May all beings be happy. May all beings be free from suffering."
- Recite *Om Mani Padme Hung* while spinning the prayer wheel.
- Visualize beams of light, bright like the sun, radiating out from the prayer wheel in all directions. The light beams destroy the negative karma and sufferings of all beings.
- Dedicate the merit of one's prayer wheel practice to purify the underlying cause of another's illness to promote healing.

The Colorful Tapestry of Prayer Flags: A Symbolic Journey of Serenity and Blessings

In the serene landscapes of Nepal, Tibet, and other Asian countries, prayer flags flutter in the wind, carrying with them the hopes, aspirations, and blessings of those who hang them. Originating from the ancient traditions of Tibet, prayer flags have become a symbol of peace, compassion, and spiritual harmony. These vibrant flags, adorned with sacred symbols and prayers, serve not only as decorative elements but also as powerful conduits for positive energy and well-being.

Figure 65 - Prayer flags by the 2059 Newman Road Parking lot. (c) Chuck Pettis

Prayer flags trace their roots to Tibetan Buddhism, where they are known as "lung ta," meaning "wind horse." Traditionally, the flags are inscribed with prayers, mantras, and auspicious symbols that are believed to be carried by the wind, spreading positive energy to the surrounding areas. The five colors used in prayer flags—blue, white, red, green, and yellow—represent the elements, and the arrangement of these colors is significant in Tibetan cosmology.

Hanging prayer flags is a symbolic gesture of fostering peace and harmony. As the wind gently carries the prayers and mantras inscribed on the flags, they spread positive intentions, creating an atmosphere of tranquility. Their presence serves as a reminder of one's spiritual journey and a connection to a higher consciousness. The fluttering flags invoke a sense of mindfulness and inner peace.

Prayer flags have the power to dispel negative energies and obstacles. The sacred symbols and mantras create a protective aura, shielding individuals and communities from harm and adversity.

Prayer wheels and prayer flags may be purchased from:

- Dharmashop: https://www.dharmashop.com/

- Etsy: https://www.etsy.com/
- Amazon: https://www.amazon.com/

"Altruistic Magic"

The Stupa, prayer wheels, and prayer flags are "altruistic magic." Each of these symbolic radiators of love and blessings is used to benefit countless others, as well as yourself. At the same time, because of the many prayers being broadcast to bring others happiness and relief of suffering, there is a karmic benefit. When you do good for others, good will come to you in the future. H.H. Dalai Lama once called this "selfish altruism." Over time, a place with these spiritual practices can become a celestial realm, a sacred ecology of peace and happiness.

Questions for Group Discussions

These questions aim to facilitate a deep and engaging discussion, encouraging participants to reflect on their personal experiences and beliefs while exploring the spiritual significance of the stupa, prayer wheels, and prayer flags.

- How do the spiritual practices described in the chapter compare to those in other religious or spiritual traditions you are familiar with? What similarities and differences can you identify?
- What resonated most with you in this chapter? Why?
- What are the reported benefits of circumambulating a stupa, as mentioned in the text? Do you believe in the transformative power of such rituals? Why or why not?
- Prayer wheels are described as powerful spiritual tools. What are the intended benefits of spinning a prayer wheel according to the chapter?
- How do you feel when you see prayer flags in your environment?
- The chapter outlines various benefits of prayer wheels, including purifying negative karma and promoting healing. How do you interpret these benefits in the context of your own spiritual or religious beliefs?

Chapter 6: The Power of Space to Alter Consciousness

Spaces, whether natural or constructed, have an extraordinary ability to influence human consciousness. This concept interweaves psychology, neuroscience, architecture, design, and ecology, demonstrating how our surroundings can shape our mental states, emotions, and perceptions.

Sensory Stimulation:

Spaces engage our senses to evoke specific emotional and psychological responses:

- **Lighting:** Natural or warm, soft lighting fosters calm, while harsh, fluorescent lighting can induce stress or anxiety.
- **Sound:** Quiet spaces or natural sounds promote relaxation and mindfulness, while noisy environments may cause overstimulation or stress.
- **Color Psychology:** Colors influence mood—blue and green are calming, while red can heighten alertness or intensity.
- **Textures and Smells:** Tactile and olfactory elements further enhance emotional connections to a space.

Architecture and Spatial Design:

The design of a space can influence how we feel, think, and behave:

- **Open vs. Confined Spaces:** Open spaces inspire freedom and creativity, whereas enclosed spaces can evoke safety or claustrophobia.
- **Flow and Movement:** Spaces that encourage wandering, like gardens or labyrinths, foster introspection and meditative states.
- **Sacred Geometry and Symbolism:** Structures such as cathedrals or temples evoke awe and transcendence, encouraging spiritual shifts in consciousness.

Psychological Anchoring:

Spaces often act as anchors for memories, emotions, and experiences:

- **Conditioned responses:** Spaces like yoga studios or meditation rooms can trigger automatic calming effects.
- **Familiarity and safety:** Returning to familiar or safe environments lowers stress, shifting the brain toward relaxation and away from fight-or-flight responses.

Nature and Biophilia:

Natural environments hold unparalleled power to shift consciousness. The concept of biophilia (our innate connection to nature) suggests that people are mentally and physically healthier when in natural settings:

- **Attention restoration theory:** Nature restores focus and reduces over-stimulation.
- **Mindfulness and presence:** Being in nature draws attention to the present moment, fostering mindfulness.
- **Transcendence and awe:** Vast landscapes, like mountains or oceans, often inspire awe, leading to feelings of interconnectedness and spirituality.

Social and Cultural Context: The power of space to alter consciousness is also shaped by social and cultural factors:

- **Cultural significance:** Pilgrimage sites, memorials, or ancient ruins evoke shared emotional and spiritual experiences.
- **Group dynamics:** Communal activities, such as festivals, religious ceremonies, or concerts, create altered states of consciousness through shared energy and heightened emotions.

Spaces Designed for Altered States:

Some spaces are purposefully created to induce shifts in consciousness:

- Meditation rooms and Zen gardens: These foster tranquility, reflection, and heightened awareness.
- Nightclubs and concert venues: Sensory elements like music, and crowds induce ecstatic or trance-like states.
- Immersive art installations: Alter perceptions of space, time, and reality for profound consciousness shifts.

Observing and Respecting Birds and Wildlife

Nature's living beings provide unique lessons in mindfulness and presence. Animals, particularly birds, are highly sensitive to human emotions, sounds, smells, and behaviors. They respond to our presence with intricate high-speed communication systems relaying signals through their calls and movements to alert others that humans are nearby.

Listening to Wildlife

When you walk outside, tune into this communication. Listen carefully to the birds, chipmunks, and other wildlife. By their calls and behavior, they reveal not only your presence but also who or what you are to others in the area.

When outside, birds and wildlife can sense our emotions, body language, and mood. If we are angry, upset, or in a bad mood, the birds will give alarm calls, letting the other birds and wildlife know that danger is coming. If a group of people are talking loudly and not paying attention to their impact on the world around them, the birds will give alarm calls. The alarm messages are passed along from bird to bird. "The human is coming!" "The human is coming!" This is called a bird plow.

If we are happy, relaxed, mindful, and walking slowly and quietly the birds will be aware of that.

Practicing Invisibility

- Indigenous traditions encourage "jungle etiquette," where quiet observation allows wildlife to acclimate to human presence.
- Avoid direct eye contact with animals, as it may be perceived as a threat. Use soft voices and respectful gestures to show kindness.

Building Connection

- If you encounter a deer, rabbit, bird, or other wildlife, take a moment to stop and respect their space. Avert your gaze and wait, showing them you mean no harm. In a soft, gentle voice, you might say, "Sorry, I didn't mean to disturb you."
- Address wildlife, trees, and plants as you would friends, fostering mutual respect. For example, greet a tree with, "Good afternoon, Sequoia."

- Naming animals you encounter regularly creates a sense of familiarity and connection.

Practice Stillness

- To see and hear more birds and wildlife, practice stillness.
- Sit quietly in one spot and let nature acclimate to your presence. After a while, you will begin to hear birdsongs again.

Minimize Impact – Leave Your Dog at Home

- If you want to deepen your connection to wildlife, consider leaving your dog at home or in your car (if the temperature permits).
- Research (https://shorturl.at/bGJ9p) has shown that the presence of dogs, even when leashed, has a significant impact on wildlife:
  - Birds and other animals often perceive dogs as predators, disrupting their normal behaviors.
  - Deer activity is significantly reduced on trails where dogs are allowed.
  - Small mammals such as squirrels, rabbits, chipmunks, and mice also exhibit reduced activity near trails frequented by dogs.
  - Dogs may displace wildlife from their habitats through chasing behavior.
  - Dogs can unintentionally spread diseases such as canine distemper, rabies, parvovirus, plague, giardia, and muscle cysts.

By respecting wildlife's need for space and tranquility, you contribute to their well-being and enrich your own experience of nature. Mindful presence, gentle observation, and acts of kindness build a deeper harmony between you and the natural world.

In short, spaces can serve as powerful tools for shifting human consciousness, whether by fostering relaxation and focus, triggering introspection, or evoking feelings of awe, wonder, or unity.

My Studies In the Power of Space to Alter Consciousness.

I earned a B.A. degree in Psychology from Carnegie-Mellon University in Pittsburgh. I started with a major in mathematics but switched to psychology when I took the Intro to Psychology class taught by the incredible and inspiring Sandra Bem and Daryl Bem.

Figure 66 - 1970 photo of Chuck Pettis in front of 3-frequency icosahedron geodesic dome. Part of a network of domes and inflatables at Carnegie-Mellon University for the first Earth Day. Photographer unknown

During my psychology studies, I was particularly influenced by the works of Edward T. Hall: *The Silent Language* (https://a.co/d/b4aGzBd) and *The Hidden Dimension* (https://a.co/d/fAuTq7G), which revealed how the relationship of oneself to others and the surroundings affect behavior and feelings.

Figure 67 - Rhombicuboctahedron Inflatable (c) Chuck Pettis

The First Earth Day

In 1970, my last year at Carnegie-Mellon, I was part of a group that built geodesic domes and inflatables for the first Earth Day. Spending time in domes and inflatables was amazing because we could all tell that we felt different in these non-rectangular structures. People were quiet, stopped, and captivated by a space that was not familiar and completely new.

Figure 68 - Circa 1969 photo of John Paskiewicz spread armed against a polyethylene pillow inflatable. (c) Chuck Pettis

My mentor at Carnegie Mellon was John Paskiewicz, a design professor who had attended Buckminster Fuller's design department at Southern Illinois University. John turned me on to Fuller and sponsored my independent study courses. I became inspired by Buckminster Fuller's vision of making the Earth work for 100% of humanity, so I studied at Buckminster Fuller's design school at Southern Illinois University at Carbondale.

The SIU Design Department was a very cool place. Al Gowan later wrote a book about it titled *Shared Vision: The Second American Bauhaus* (https://a.co/d/9MzeG3L) of which I am proud to be a part.

In my first year at the Design Department, I built a 5/8's 3-frequency geodesic dome that featured great circle arcs to the south, so when we had full moon meditations, the moon would move through these arcs. More importantly, people became very silent in the dome.

Figure 69 - Chuck Pettis beginning the assembly of the Carbondale geodesic dome in 1971. Photographer unknown.

The acoustics were unusual. As we sat cross-legged around the circular interior, the sound was amplified. Because of that, we became silent. Without the corners of a rectangular building, the eye adjusts to the spherical shape of triangles. Everyone becomes more mindful. The space itself altered people's consciousness, and silence was pervasive.

Figure 70 - 5/8's 3-frequency icosahedron dome built in Carbondale, IL. (c) Chuck Pettis

When we enter a space or place we have not been in before and do not have memories and expectations, we temporarily forget our perceptual baggage. Unfamiliarity is a setup for a digital uplift in consciousness. Rather than a slow continuous growth (analog) process, a digital uplift is a jump up, an instant change in perception. By disabling our embedded way of looking at things, we can see things as if for the first time, and the potential for awe and a sense of wonder, which abides nowhere else but, in the present, can emerge.

By removing the filters of our conditioned perceptions, our doubting mind can be subdued, enabling transcendental and transformative experiences. For example, I met a woman at Earth Sanctuary who told me, "I am someone who is always looking for deficiencies and what could be improved. At Earth Sanctuary I see nothing that can be improved."

At Earth Sanctuary, visitors experience a variety of special spaces that many people have never experienced before. The dolmen, the labyrinth, the stone circles, the medicine wheels, the stupa, the ley line sculpture—all of these put visitors in a new kind of space that enables them to have transformational spiritual experiences, a heightening of consciousness and spiritual awareness.

The Cottonwood Stone Circle was designed to be a place for prayer, meditation, and spiritual discussion. Because of the many blessings bestowed here by spiritual leaders, visitors, and retreatants, the Cottonwood stone circle has a spiritual power that can help you achieve realizations. Put your hands on the stones and fill them with your love and blessings or if you're experiencing grief or sadness, the stones may relieve you of grief and sadness.

The Earth Sanctuary Labyrinth was constructed as a tool for walking meditation, with the idea that walking into the labyrinth's center and out again can be experienced as a metaphor for life's journey. *The walk through the labyrinth walked me back through time to let the past go. The journey out gave me hope and direction for the future.* The labyrinth is also a place to be aware of the landscape Angel, the Devas, and the nature spirits that are all around. Send them love and appreciation for the great work they do.

The Medicine Wheel, one of the most powerful and cherished symbols of Native American belief, is an ancient place of prayer sacred to Native Americans, symbolizing the totality of existence. The Earth Sanctuary Medicine Wheel is a sacred space designed to amplify the power of your prayers and connect you with all the Grandmothers, Grandfathers, Mother Earth, the Moon, the Sun, the Sky, and the Creator, on the eternal circle of life.

The Stupa is a place of great spiritual power. Seven Tibetan Buddhist ceremonies were done here for the benefit of all beings. Circumambulating the stupa enables your prayers to be amplified to help those in need of happiness and relief of suffering. It is a holy place and a place of reverence.

The Dolmen is a place for meditation and reflection. The millions of prayers and mantras done here make it the perfect place to clear your mind and discover an intense, unmistakable connection with the earth. *While doing some quiet chanting yesterday in the Earth Sanctuary's spectacular Dolmen, a hummingbird came into the space and hovered for a long time like a blessings to my meditation. What an incredible sanctuary this is that even the birds feel safe and happy to join in the space!*

The Earth Sanctuary Fen Stone Circle is a place for compassion, love, and blessings. *We found the energy in the Fen Stone Circle to be very powerful. We are in appreciation of the dedication that has been gifted to this place. It is very evident that the earth here is holding the loving energy for all that visit and for all mankind.*

The Compressed Cube Tensegrity Sculpture consists of four long aluminum tubes connected with cables arranged in such a way that the metal tubes do not touch each other. The sculpture appears to be floating in space at the same time it is locked into form. This presents a moment of stasis. It is a perfect metaphor for life demonstrating that we can balance the things in life pulling and pushing us to create equanimity.

The Ley Line Sculpture is about energy and the effect of energy on your consciousness. By feeling the sensations of ley lines and a ley line power center, you will be able to recognize that feeling if you visit a place of power, such as an ancient monument or a cathedral. These earth energies create an atmosphere around a place. Because these places hold the strong emotions of visitors, you can infuse a sacred space with your love, prayers, and blessings.

What special places have you been to that altered your consciousness? If you could design your perfect atmosphere, what would it feel like?

Questions for Group Discussions

These questions can help facilitate a deeper discussion on how different spaces influence consciousness and the transformative power of specific environments.

- Have you ever been in a space that profoundly affected your mood or consciousness? What was it about the space that made you feel that way?
- The author was inspired by Edward T. Hall's work on how space affects behavior and feelings. Can you think of other examples or studies that show the impact of space on human psychology?
- How do you think the design and architecture of a space can influence social interactions and personal well-being?
- The author mentions feeling different in non-rectangular structures like geodesic domes. Why do you think these shapes might alter consciousness compared to traditional rectangular buildings?
- Have you experienced being in a geodesic dome or a similar structure? If so, how did it feel differently from other spaces?
- The author talks about unfamiliar spaces leading to a "digital uplift in consciousness." What do you think this means, and how can unfamiliarity with a space lead to such an uplift?
- Do you think it's possible to design urban environments that promote similar transformative experiences? How might that be achieved?
- What role do you think cultural and historical contexts play in the design and impact of spiritual spaces?
- The author suggests that natural spaces can enhance spiritual awareness. How do you think being in nature differs from being in man-made environments in terms of affecting consciousness?

Chapter 7:
Connect and Communicate with Trees

For centuries, humans have viewed trees as mere lifeless objects, providing shade and oxygen, but with no cognitive abilities or communication skills. However, recent scientific research has revealed that trees possess intelligence and can communicate with each other and even with other species.

What does it mean that trees are intelligent? Trees can identify, locate, and allocate resources; communicate and collaborate with other organisms; compete; and make choices to survive and grow.

Trees are aware of themselves and their surroundings:

1. By perceiving the scent that other trees release;
2. By tasting the nutrients that the roots find in their exploration for food;
3. By tasting the saliva of insects and sending out pheromones that summon beneficial predators;
4. By perceiving the moisture of the soil;
5. By feeling the touch and injuries from animals and insects; and
6. By constantly adjusting to changes in temperature and sunlight.

Figure 71 - Intelligent Trees DVD cover (https://shorturl.at/onGou)

The Mycorrhizal Network: the Wood Wide Web

When you walk the trails of Earth Sanctuary, imagine you are walking through a neighborhood of living beings called trees, plants, and fungi. They are all connected, not through phone lines or the internet, but rather through the mycorrhizal network, the "wood wide web." Through the mycorrhizal network, life-giving nutrients are shared to keep the forests healthy, and messages and signals are sent in case of attack by insects or an injury. Mother trees care for their offspring by sharing nutrients to keep the young trees healthy.

Figure 72 - An illustration of the mycorrhizal network beneath the forest floor, showcasing the interconnected "wood wide web" that supports and nourishes the forest ecosystem By DALL-E.

Forests are "social super organisms" connected underground through their root systems and mycelium. Through this "wood wide web" trees and fungi have a symbiotic relationship—fungi exchange nutrients and water for sugars produced by the tree's photosynthesis. The network of underground fungi enables trees to altruistically communicate, share resources, warn one another of potential threats such as pests and drought, and even transfer carbon to support neighboring trees.

Figure 73 - The Trees of Earth Sanctuary (c) Michael Stadler and Chuck Pettis

Tree Brains

Trees have "brains" that perceive, communicate, and remember, just like animals. However, trees can't move like animals who require a "portable" brain to find food and shelter. Trees don't need a compact and portable brain because food comes to the tree through its leaves and roots.

One of the ways that trees communicate is through electrical impulses. In trees, electrical signals can travel at rates ranging from fractions of an inch per second to a few inches per second. This is significantly slower than the speed of nerve impulses in humans which can vary from 1-100 meters/second, over two hundred miles per hour.

Research has shown that trees are capable of emitting volatile organic compounds (VOCs), which act as chemical signals. These VOCs can attract beneficial insects and repel harmful ones, as well as communicate with other trees about potential threats or environmental changes.

Trees Can Remember

Additionally, studies have shown that trees can remember past experiences and respond to them accordingly. For example, some trees can differentiate between the smell of insects that have previously attacked them and those that have not and will release specific VOCs to attract predators to eliminate the threat.

Whereas animals can move quickly, trees move very slowly. This isn't necessarily a disadvantage. At Earth Sanctuary, most trees will live to 250, 500, even thousands of years old. One of the oldest trees on Earth is a spruce in Sweden that is more than 9,500 years old! (We have planted many spruce trees at Earth Sanctuary.)

Even though they move slowly, trees can grow based on their environment and surrounding trees. For instance, in crowded forests, trees will often grow taller and thinner to compete for sunlight, while in open spaces, they will grow broader and thicker to capture more light. Trees can grow new branches or shed unneeded ones to optimize light capture. There is a spruce at Earth Sanctuary that I observed after learning this. Its sunny south side is noticeably denser with branches than its shady north side.

In light of these discoveries, scientists are beginning to recognize the intelligence of trees and the important role they play in ecosystems. As humans continue to exploit natural resources and alter landscapes, it is crucial to understand and appreciate the complex communication and adaptive abilities of trees to ensure their survival and that of the ecosystems they support.

In conclusion, trees are much more than just objects in our surroundings. They possess remarkable intelligence and the ability to communicate through underground networks, chemical signals, and growth patterns. As we strive to preserve our natural world, it is important to recognize and appreciate the intelligence of trees and the vital role they play in maintaining the health of our planet.

When walking in a forest, know that you are walking in a living, aware, and intelligent community of trees. Come and immerse yourself in the healing forest community of Earth Sanctuary or a forest near you. Take in the forest atmosphere and experience forest bathing. Spending time in a forest reduces stress, anxiety, and anger, strengthens the immune system, and promotes overall health and well-being.

For more information on the intelligence of trees, I recommend:

- The Mind of Plants - Narratives of Vegetal Intelligence (https://a.co/d/163VYFq). Edited by John C. Ryan, Patricia Vieira, and Monica Gagliano.
- The Hidden Life of Trees; What They Feel, How They Communicate (https://a.co/d/2SaxLcX) by Peter Whohlleben

How to Communicate with Trees

One simple way we can connect, communicate, and relate to trees is to breathe on their leaves. When you breathe near the leaves or needles of a tree, your carbon dioxide is absorbed by the tree, helping with its photosynthesis. The conductivity of the leaf changes; the tree is sensing you!

In the 1970s, when I was living in Ithaca, New York, the founders of the intentional community called Findhorn—Dorothy MacLean, Peter and Eileen Caddy—visited the Foundation of Light several times. At the time, the Foundation of Light was a school where we studied ancient wisdom. John and Kate Payne, as well as healers from the U.S. and England, taught touch and light healing techniques, dowsing, meditation, and more.

Dorothy could communicate with the intelligence of Nature through the Devic or angelic realms. By contacting and cooperating with the devas of the plants in their garden, she was able to create an abundant garden in the Scottish sand landscape, including giant cabbages and winter-flowering roses that became well publicized at the time. (For more information on Devas, I recommend the book, Call of the Trees (https://a.co/d/7hNLoX5), by Dorothy MacLean.)

On one visit to the Foundation of Light in the mid-1970s, Dorothy led a day-long workshop on communicating with Devas that resonated with me, and I found that I, too, could connect and communicate with Devas. In Chapter Three I have already explained, in essence, the spiritual practice that Dorothy taught on how to contact and communicate with Devas.

Large Trees Are Essential for the Well-Being of the Earth

Dorothy was one of the first to warn about Earth's environmental problems and the importance of planting trees. On another occasion in Ithaca,

I attended a talk by Dorothy on the importance of large, elderly trees to the planet. She said the trees were calling on us to protect old-growth trees and to re-forest the Earth. Large trees are essential for the well-being of the Earth. So, love trees and plant them with love! I have never forgotten that talk. That talk motivated me to re-forest Earth Sanctuary with a 500-year plan to create an old-growth forest.

Why old growth? The tree Devas says, "We need a fully grown tree for our real influence — a child cannot do what a man can do — but we think that, with our concentration and your co-operation, we can make the influence strong soon[9]." The Earth spirits want us to preserve old-growth trees and plant as many long-life-growing trees as possible.

"How special and important to be in a forest that exists for the sake of the forest, itself."

To date at Earth Sanctuary, we have planted over 3,350 trees. When I purchased the Earth Sanctuary property in 2000, there were only 8 species of trees. Now there are 35 tree species. Here is a link to a presentation on The Trees of Earth Sanctuary (https://shorturl.at/q998n).

At Earth Sanctuary, the Landscape Angel and Devas oversee and help the work we are doing to increase the biodiversity of trees and plants and to create an old-growth forest for the future. If you sense their presence while visiting Earth Sanctuary, send them love and gratitude for her help in making Earth Sanctuary such a peaceful and magical place.

Nurturing Nature: The Case for Assisted Migration of Trees

In the face of rapid climate change and its undeniable impact on ecosystems worldwide, scientists and conservationists are exploring innovative strategies to help species adapt and thrive. Assisted migration, a concept gaining traction in the realm of environmental conservation, involves the deliberate movement of species to new, more suitable habitats. Among the myriad species facing the brunt of climate change, trees play a pivotal role in maintaining ecological balance and supporting diverse forms of life. Assisted migration of trees is emerging as a promising approach to ensure the resilience and survival of these vital organisms.

Assisted migration, also known as managed relocation or translocation, involves the controlled movement of species beyond their natural range to regions where they can better cope with changing environmental conditions. In the context of trees, this includes transplanting seedlings or mature trees to areas with more favorable climates or soil conditions. The goal is to facilitate the migration of species that might otherwise struggle to adapt to their changing native habitats.

My wife, Claudia, grew up in Seattle. She remembers that it always rained in June and that it rained on most July 4 events. It also rained regularly in September. Not so today. When I purchased Earth Sanctuary in 2000, we started planting trees and native plants, as part of my 500-year plan to return Earth Sanctuary to an old-growth forest. It was immediately apparent that the lack of rainfall from May to September with hot summers was too much for many native plants. So, we installed a watering system to nurture the native plants and put up fencing to prevent the rabbits and deer from eating all the new "yummy" fertilized plants.

As global temperatures rise and precipitation patterns shift, many tree species face challenges such as increased susceptibility to pests, diseases, and changing soil conditions. Assisted migration can help trees move towards more suitable environments, allowing them to thrive rather than succumb to adverse conditions.

Between 2008 and 2009, we planted 1,300 trees on the east side of Earth Sanctuary to create an arboretum. Over 300 trees did not make it. The trees that survived were drought-resistant trees from habitats east and south of Puget Sound—Western White Pines, Ponderosa Pines, Oregon Ash, Sequoias, Redwoods, and other similar species. Climate change is a reality when it comes to trees.

We need trees!

Trees provide essential habitats for countless other species. By ensuring the survival of tree species through assisted migration, we contribute to the preservation of entire ecosystems. This, in turn, helps maintain biodiversity and the intricate web of life that depends on these towering organisms.

Trees play a crucial role in providing ecosystem services, including carbon sequestration, oxygen production, and soil stabilization. Assisted migration can contribute to the continuity of these services, mitigating the impact of climate change on the overall health of ecosystems.

As the effects of climate change become increasingly apparent, the need for proactive and innovative conservation strategies grows more urgent. Assisted migration of trees represents a potential tool in the conservation toolkit, offering a way to help vulnerable species navigate the challenges posed by a rapidly changing environment. However, careful consideration, research, and ethical deliberation are paramount to ensure that assisted migration efforts contribute positively to the well-being of both trees and the ecosystems they inhabit.

As we venture into this new frontier of conservation, the delicate dance between human intervention and the inherent resilience of nature must be approached with humility and a deep understanding of ecological intricacies.

The Intelligence of Trees

In forests deep, where ancient secrets hide,
Resides a wisdom known to ancient trees.
Their branches reaching out, spreading wide.
They whisper tales carried upon the breeze.

Though rooted to the earth, they yearn to learn,
Absorbing knowledge from the world around.
Their silent minds in harmony discern
The subtle truths within the forest's sound.

With patience born of fleeting years,
They witness seasons come and seasons go.
Each ring a testament to life's frontier,
Inscribed with stories only they can know.

So let us ponder on their silent speech,
For trees, dear friend, hold nature's wisdom's reach.

Questions for Group Discussions

These questions aim to foster a deeper understanding and appreciation of the intelligence and importance of trees, encouraging thoughtful reflection and discussion on how we can better coexist with and support our natural environment.

- What aspects of tree intelligence do you find most fascinating (e.g., sensing surroundings, communicating with other trees, remembering past experiences)?
- How does the concept of the "wood wide web" change your understanding of forest ecosystems?
- Have you ever felt a spiritual or emotional connection with a tree or a forest? Can you describe that experience?

- Do you believe that communicating with trees, as suggested by the author, can have a real impact on your well-being? Why or why not?
- What are your thoughts on the concept of assisted migration for trees as a response to climate change? What potential benefits and challenges do you see in implementing assisted migration on a large scale?
- After reading about the intelligence of trees, do you feel inspired to visit or engage with a forest or natural area differently? How?

Chapter 8:
The Earth Sanctuary Memorial Tree Program

Figure 74 - The spiritual light in the forest. (c) Michael Stadler

The primary goals of the Earth Sanctuary Memorial Tree Program are to provide comfort and healing to those who have lost a loved one and to enable individuals to select their final resting place.

The Earth Sanctuary Memorial Tree Program started in early 2020. In the previous two years, three people had asked to have a memorial for a loved one at Earth Sanctuary. I thought to myself, if people see Earth Sanctuary as a trusted place for a memorial, why not offer trees as memorials to help fund our tree planting and maintenance program? The program was an immediate success, and we now have over one hundred Memorial Trees at Earth Sanctuary.

Figure 75 -A granite sculpture and a prayer wheel were two memorials that kickstarted the Memorial Tree Program

Reasons to Honor the Life of a Loved One With a Memorial Tree at Earth Sanctuary

Have you experienced the death or loss of a loved one? Do you feel sad and helpless? Find comfort and healing by creating a living legacy that honors someone who has died. The Earth Sanctuary Memorial Tree program establishes a special tree where one can remember, grieve, pray, and commune in memory of a deceased loved one in a natural, peaceful, and beautiful setting.

A Memorial Tree is a much less expensive alternative to placing remains in a cemetery and purchasing a tombstone For those choosing cremation, aquamation, or recomposition (the composting of human remains), a Memorial Tree can be a natural, living, and sustainable alternative to cemeteries, tombstones, and mausoleums.

A Memorial Tree is good for the environment. Its location is natural, peaceful, and beautiful, and the cost is less than a traditional graveyard. And you are helping Earth Sanctuary achieve its 500-year plan to return the property to a mature old-growth forest.

Figure 76 - Beverly Rose by her Western White Pine Memorial Tree (c) Chuck Pettis

Select Your Memorial Tree While You Are Still Alive

- Choose your final resting place: a tree that will live for hundreds, even thousands of years.
- A natural, living, and affordable alternative to placing your remains in cemeteries or mausoleums
- So much better than a tombstone or grave marker!
- Bury your cremated remains by your Memorial Tree.

Ready to Order an Earth Sanctuary Memorial Tree?

- Visit the Earth Sanctuary Memorial Tree website (https://earthsanctuary.org/memorial-tree-program/).
- Contact Earth Sanctuary Owner and Creator Chuck Pettis at cpettis@earthsanctuary.org.

Chapter 9:
The Power of a Diverse and Healthy Ecosystem

The Earth Sanctuary 500-Year Plan to Create an Old-Growth Forest

Figure 77 - Aerial view of forested Earth Sanctuary looking Southwest over the Fen, Middle, and West Ponds. Photo source unknown.

The landscape and ecological restoration design currently underway at Earth Sanctuary was born from my vision to create a sanctuary for both nature and people. The Earth Sanctuary's 72 acres are unique in the South Whidbey Island landscape for its complexity and abundance of aquatic, wetland, and forest habitats.

Figure 78 - The Fen Pond from the bench on the East side of the Fen Pond. (c) Chuck Pettis

"I think having land and not ruining it is the most beautiful art that anybody could ever want." – Andy Warhol

I've been in the marketing and branding business my entire career. I have done many branding projects with non-profit environmental organizations (https://brand-solutions.com/clients.html) to help them protect the environment for future generations. Rather than worry about all of Earth's environmental problems that I have no control over, I decided to take personal responsibility and enhance biodiversity by creating and executing a 500-year plan to create an old-growth forest at Earth Sanctuary, while at the same time, creating inspiring and unique artworks for personal renewal and spiritual growth.

Nature Is The Top Priority

Earth Sanctuary is a place where Nature is the top priority. It is a place to see, feel, smell, touch, and hear nature. As you walk around Earth Sanctuary, know that we are combining the state-of-the art in ecological restoration, with art and spirit to create a peaceful and magical place for your personal enjoyment, healing, and blessing. Your visitor fees enable me to keep Earth Sanctuary open to the public and also pay for our ongoing tree planting and maintenance. Thank you to everyone who has contributed!

The Earth Sanctuary's three ponds, known locally as the Newman Ponds, have been recognized for their importance as waterfowl habitat and are designated as a Habitat of Local Importance by the Whidbey Audubon Society and the Island County Critical Areas Program.

Ponds attract many birds. On any given spring morning one may see the local pair of nesting Ospreys and their new fledglings, adolescent and mature Bald Eagles, Wood Ducks, Hooded Mergansers, Red-winged Blackbirds, Killdeers, Tree- and Violet-green swallows, Cinnamon Teals, Great Blue Herons, among others.

Assembling a World-Class Team of Ecological Experts

Right after I purchased the 72-acre Earth Sanctuary property, I assembled a world-class team of experts in ecology, restoration, landscape design, forest planners, and feng shui. I intended to combine ecology, art, and spirituality in the design of the Earth Sanctuary landscape.

The interdisciplinary Earth Sanctuary design team included: myself, architect David Rousseau, landscape architect Dan Borroff, ecologist Kevin Fetherston, artist Mike Sweeney, landscaper Dave Schmidt, and the late spiritual advisor H.H. J.D. Sakya, head of the Sakya Monastery of Tibetan Buddhism.

The human-nature interface of both the designed and natural landscape has been deliberately cultivated from an ecological perspective. The design of the Earth Sanctuary – including stone megalithic artworks, landscape design, and the ecological restoration of historically degraded forest and wetland ecosystems – has been guided by the overall goal of preserving, protecting, and restoring the land's biodiversity.

The goals of the 500-year plan:

1. Use an exemplary and comprehensive ecological design process
2. Create a nature reserve with maximum wildlife diversity and population
3. Restore the land to mature old-growth forest with minimum impact to the environment
4. Benefit the local community economically and ecologically
5. Protect the wetlands

These qualities are the targets of the Earth Sanctuary restoration.

Figure 79 - Kevin Fetherston (left) was the primary leader of the 500-year plan and the Design with Nature methodology. (c) Chuck Pettis

The Design with Nature Methodology

At Earth Sanctuary, I wanted to "Design with Nature." When I researched how to do this, I couldn't find any "how-to" resources. So, Kevin Fetherston and my team developed an ecosystem-based approach for ecological design and development, i.e., let the land speak to us and tell us what to do, to "truly" design with nature.

1. Conduct baseline ecosystem surveys of existing conditions, including wetlands, streams, vegetation, fish and wildlife, and the environmental regulatory landscape (e.g., critical area regulatory buffers).
2. Conduct a reference ecosystem survey of South Whidbey State Park, the only old-growth forest still intact on Whidbey Island, to serve as the basis for ecological restoration design and landscaping.
3. Develop a Geographic Information System (GIS) map of the ponds, streams, and wetlands for design visualization, data management, and environmental education.
4. Develop the ecological restoration design and plan based on the baseline ecosystem survey and reference ecosystems survey.
5. Design of site roads and paths.

A 500-year plan acknowledges that the ecological communities of the Earth Sanctuary will take hundreds of years to recover from past land use practices before they develop into their fullest expression as a diverse and mature ecosystem. Also, the expression "500-year plan" acknowledges the human commitment necessary to preserve, protect, and restore the Earth Sanctuary's landscape over the long term.

Definition of "Old-growth Forest"

What exactly is an old-growth forest? Franklin et al. (1981) describe the characteristics of old-growth forests:

1. Approximately 175 to 250 years are required to develop old-growth forests under natural conditions in both Coast and Cascade Ranges. Development of old growth is faster on good sites than on poor sites.
2. Few plant or animal species are solely confined to old-growth forests, although many species— including several vertebrates, saprophytic plants, and epiphytic lichens—find optimum habitats in such forests. Some organisms, however, may require old growth to maintain viable populations. Moreover, there are substantial differences in composition and relative abundance of species between young- and old-growth forests.
3. Gross productivity is maintained at high levels in most old-growth stands, but mortality generally balances growth. Total organic matter keeps increasing because of accumulated masses of dead tree boles, mostly as down logs. (As you walk the Earth Sanctuary trails note the downed trees along the trails and in the forest.)
4. Old-growth forests are highly retentive of nutrients; large amounts are incorporated into living and dead organic matter. Losses of limiting nutrients, such as nitrogen, are low.
5. Nitrogen-fixing epiphytes are abundant in old-growth trees, and bacterial nitrogen fixation appears to be common in the large woody debris characteristic of old-growth forests.
6. The structure of old-growth forest is more heterogenous than that of young forests; coefficients of variation in tree sizes are greater, and understory patchiness is much higher than in young-growth stands.
7. Most of the distinctive features of old-growth forests can be related to four structural features; large, live old-growth trees, large snags, large logs on land and in streams.
8. A large, old growth Douglas fir is individualistic and commonly has an irregularly arranged, large, coarse branch system, and often, a long

crown. It is ideal habitat for specialized vertebrates, such as the red tree vole, Northern Spotted Owl, and northern flying squirrel, as well as nitrogen-fixing lichens.

9. Large snags are valuable as habitat for a variety of vertebrates and invertebrates and as a future source of logs.
10. Logs on the forest floor are important habitats for small mammals, including species that disperse spores of mycorrhiza-forming fungi. They also are sites for substantial bacterial nitrogen fixation and are essential as seedbeds for some trees and shrubs.

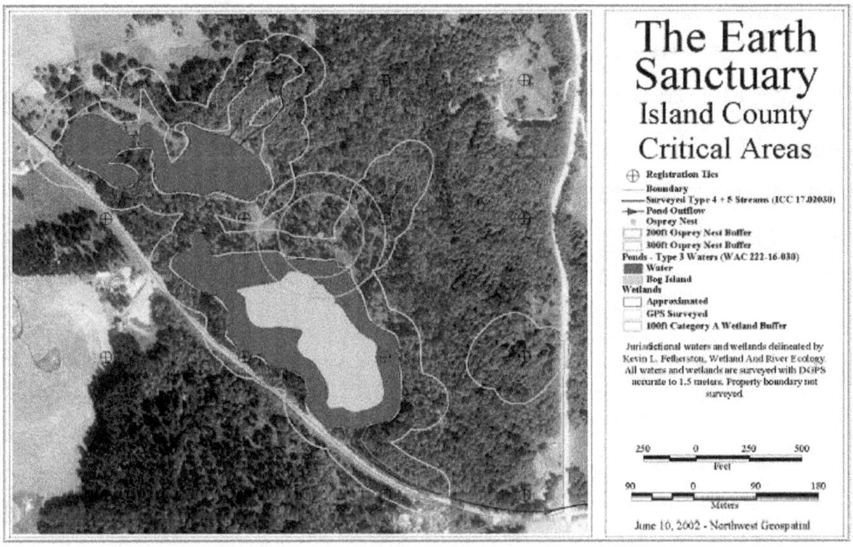

Figure 80 - Earth Sanctuary Wetland boundaries. (c) Northwest Geospatial

Mapping the Wetlands and Ponds

Before developing the landscape, restoration, and environmental artwork designs, we conducted a two-phase delineation and mapping of the Earth Sanctuary wetlands, streams, and ponds and their regulatory buffers - environmentally critical areas regulated under Island County, Washington State, and federal environmental ordinances. The resulting Island County Critical Areas map has served as a guideline for developing the Earth Sanctuary master plan including the location of landscape design elements - trail locations, landscape plantings, parking, road access- and environmental artworks.

Earth Sanctuary GPS surveys and GIS analysis and mapping were conducted by Phil Hurvitz and Luke Rogers of Northwest Geospatial LLP. Surveys of the Earth Sanctuary Critical Area boundaries and other environmental features (trails, environmental artwork locations, habitat elements) were conducted with a Trimble GPS Pathfinder ProXR (Trimble, Inc.) and Criterion 400 Laser (Laser Technology, Inc.). Given the relatively remote terrain of the Earth Sanctuary, the Trimble GPS (global positioning system) combined with Criterion Laser (hand-held rapid total survey station) allowed for a low-cost survey accurate enough for our landscape design needs (± 1.5 meters using differential GPS). Our survey approach combined both GPS-located intermediate benchmarks and traverse surveys tied into located benchmarks.

Earth Sanctuary geographic data – environmental surveys, archival aerial photography, and USGS 10m digital elevation map -- were input to ArcView (ESRI) GIS for analysis and map production.

Dan Borroff, Earth Sanctuary's landscape architect, used Vectorworks (Nemetschek North America) computer-aided design (CAD) software to create the landscape design drawings. ArcView GIS maps were imported to the CAD environment and served as base maps for Dan's landscape design work.

"A thing is right when it tends to preserve the integrity, stability, and beauty of the biotic community. It is wrong when it tends otherwise."
—Aldo Leopold, 1949, A Sand County Almanac

Goals of the Earth Sanctuary Ecological Design

- Preserve the integrity, stability, and beauty of the biotic community.
- Create an environment that allows people to engage with the Earth Sanctuary's natural communities while protecting and restoring the land's biodiversity.

To accomplish these goals Earth Sanctuary's ecological design is focused on five broad areas:

1. Artworks and landscape design
2. Wildlife habitat enhancement
3. Forest restoration
4. Riparian vegetation restoration
5. Newman Road pond/forest buffer

Strategies for wildlife habitat enhancement at the Earth Sanctuary include (Kelsey 2001):

1. Silvicultural interventions – Red alder thinning and conifer underplanting in red alder/salmonberry-dominated forest stands to accelerate the growth rate of existing understory conifers and increase forest structural complexity. Felled red alder (and other tree species) are left on the forest floor, providing cover and insects for wildlife as well as returning nutrients to the soil.
2. Preservation and addition of tree snags (i.e., standing dead trees). Snags are a major component of wildlife habitat within the forest.
3. Replacement of invasive non-native plant species (English Holly, English Ivy, reed canary grass, Himalayan blackberries) with native vegetation. The reintroduction of native plants increases forest understory diversity and provides berries, nesting materials, and cover for birds and mammals.
4. Installation of bird nesting boxes. Twenty bird-nesting boxes were installed in the Spring of 2002: (for Wood Ducks, kestrels, bluebirds, chickadees, owls, and woodpeckers). We are currently maintaining and monitoring seven nesting boxes for Wood Ducks and Hooded Mergansers, both cavity-nesting ducks who would normally have nests in old-growth tree cavities created by Pileated Woodpeckers. There is also one bat box in the forest.

We developed separate plans for riparian vegetation restoration around the three ponds, native plant restoration, and forest restoration, all permitted through the Island County Planning Department. To date, we have planted over 15,000 native plants, and over 80 species. In 2000, there were eight species of trees on the property. As of 2025, we have planted over 3,350 trees, 35 species.

Taxonomic nomenclature of species initially found at the Earth Sanctuary

Trees	**Grand fir**	**Abies grandis**
	Big-leaf maple	*Acer macrophyllum*
	Red alder	*Alnus rubra*
	Sitka spruce	*Picea sitchensis*
	Douglas fir	*Pseudotsuga menziesii*
	Willow	*Salix spp.*
	Western red cedar	*Thuja plicata*
	Western hemlock	*Tsuga heterophylla*

Shrubs	**Oregon grape**	**Berberis nervosa**
	Salal	*Gaultheria shallon*
	Holly	*Ilex verticillatus*
	Currant, Gooseberry	*Ribes lacustre*
	Wild rose	*Rosa gymnocarpa*
	Thimbleberry	*Rubus parviflorus*
	Himalayan blackberry	*Rubus procerus*
	Salmonberry	*Rubus spectabilis*
	Evergreen blackberry	*Rubus laciniatus*
	Trailing blackberry	*Rubus ursinus*
	Elderberry	*Sambucus racemosa*
	Huckleberry	*Vaccinium ovatum*
	Red huckleberry	*Vaccinium parvifolium*
Herbs	**Colonial bentgrass**	**Agrostis capillaris**
	Dwarf mistletow	*Arceuthobium spp*
	Unidentified sedge	*Carex spp.*
	Bedstraw, Cleavers	*Galium triflorum*
	Small-fruited Alumroot	*Heuchera micrantha*
	Iris	*Iris spp.*
	Twinflower	*Linnaea borealis*
	Skunk cabbage	*Lysichiton americanum*
	False lily of the valley, bead ruby	*Maianthemum dentatum*
	Mitrewort, Bishop's cap	*Mitella pentandra*
	Miner's lettuce	*Montia sibirica*
	Water-parsley	*Oenanthe sarmentosa*
	Grass	*Unidentified Poaceae*
	Chickweed	*Stellaria media*
	Coolwort, Foamflower	*Tiarella trifoliata*
	Starflower	*Trientalis latifolia*
	Stinging nettle	*Urtica dioica*
Ferns and fern allies	**Horsetail; Scouring rush**	**Equisetum arvense**
	Horsetail; Scouring rush	*Equisetum telmateia*
	Lady-fern	*Athyrium filix-femina*
	Deer fern	*Blechnum spicant*
	Shield-fern	*Dryopteris expansa*
	Wood-fern	*Dryopteris gymnocarpa*
	Oak-fern	*Gymnocarpium dryopteris*
	Sword fern	*Polystichum munitum*
	Bracken fern	*Pteridium aquilinum*

Lichens	Lichen	Lichen
	Lichen	*Unidentified lichen #1*
	Usnea lichen	*Usnea filipendula*

When we started planting, we found that due to three months of no rain in the summer, a watering system was needed, so we installed a half-mile water line and a drip irrigation system. The rabbits and deer immediately started eating all the new plants, so we had to install fencing.

We did tests and found that MycoGrow from Fungi Perfecti (unfortunately MycoGrow is no longer available) under the newly planted trees and plants greatly extends and enhances tree and plant nutrient-gathering abilities and also helps protect the trees and plants from disease.

Oh, and the blackberries. When I first acquired Earth Sanctuary, you needed a machete to go anywhere. The property was completely overgrown, mainly with acres and acres of blackberries. We started removing them by hand, then moved to hedge trimmers, weed eaters, and a munching machine that turns blackberries into mulch. Note: the blackberries grow back unless you remove the root balls or regularly cut them back.

The Earth Sanctuary Ponds

Figure 81- The Middle Pond (c) Chuck Pettis

The West and Middle Ponds were designed, circa 1970, by the US Department of Agriculture Soil Conservation Service (SCS) for waterfowl and aquatic wildlife habitat. The natural stream courses that enter the Middle Pond historically flowed down a gradient through the areas now occupied by both the Middle and West Ponds and then off-property alongside Newman Road. The western end of the Middle Pond was excavated, and a levy and cement box weir (i.e., water control structure) was constructed. The weir dammed the flow of the streams and water backed up throughout the area creating the Middle Pond. The extensive number of dead-standing and fallen trees throughout the Middle Pond is evidence of the now-flooded historic forest.

In the late 1950s, the West Pond was excavated to a depth of 10-12 feet and surrounded by an artificial earthen berm. An additional box weir at the southwestern edge of the West Pond controls its water level and outflow. The result of both pond creations is that water flows from the hill slope streams to the Middle Pond over its weir into the West Pond and during times of high water flows off-site to the drainage ditch running parallel to Newman Road.

The Fen Pond – An Extraordinary Ecosystem

Figure 82 - The Fen Lookout with Peace Pol and bench. (c) Chuck Pettis

The Fen Pond, a spring-fed peatland, is an extraordinary ecosystem, much different than the forests of Earth Sanctuary. The Earth Sanctuary plant species inventory of the Fen, as of Spring 2002, included: seventy-one species of vascular plants (see table below), 5-10 species of Sphagnum mosses, and 10-20 species of nonsphagnous mosses and liverworts. A number of these plant species were found exclusively in the acidic bog environment. These included cotton grass (Eriophorum chamissonis), sundew (Drosera rotundifolia), bog cranberry (Vaccinium oxycoccos), bog laurel (Kalmia microphylla), Labrador tea (Ledum groenlandicum), and Sphagnum mosses. In addition to the plant species, 10-20 species of lichens were noted (i.e., lichens are fungal/algal or fungal/bacterial symbioses in the Kingdom Fungi). While this is not a complete inventory of plant species occurring at the Earth Sanctuary fen/bog, it certainly includes the dominant species that define vegetation composition and structure. No rare, threatened, or endangered species have been noted to date.

Table 1. Flora Earth Sanctuary Fen/Bog Ecosystem
(Nomenclature follows Hitchcock and Cronquist 1998 for vascular plants, McCune and Geiser 1997 for lichens, and Lawton, 1971 for mosses)

Group	Common Name	Taxonomy	Growth Form
Vascular Plants	Duckweed	*Lemna minor* L.	Aquatic herb
	Eurasian water-milfoil	*Myriophyllum spicatum* L.	Aquatic herb
	St. John's wort	*Hypericum anagalloides* C. & S.	Aquatic herb
	Bracken fern	*Pteridium aqulinum* (L.) Kuhn	Herb
	Bur-reed	*Sparganium* spp.	Herb
	Cat-tail	*Typha latifolia* L.	Herb
	Common rush	*Juncus effusus* L.	Herb
	Cotton-grass	*Eriophorum chamissonis* C.A. Mey	Herb
	Hard-stemmed bulrush	*Scirpus acutis* Muhl.	Herb
	Marsh cinquefoil	*Potentilla palustris* (L.) Scop.	Herb
	Purple-leaved willowherb	*Epilobium ciliatum*	Herb
	Skunk cabbage	*Lysichitum americanum* Hultén & St.John	Herb
	Smooth sedge	*Carex laeviculmis* Meinsch.	Herb
	Sundew	*Drosera rotundifolia* L.	Herb
	Undetermined Grass	Poaceae	Herb
	Yellow water lily	*Nuphar polysepalum* Engelm.	Herb
	Bog cranberry	*Vaccinnium oxycoccos* L.	Shrub
	Bog laurel	*Kalmia microphylla* (Hook.) Keller	Shrub
	Hardhack	*Spiraea douglasii* Hook.	Shrub
	Labrador tea	*Ledum groenlandicum* Oeder	Shrub
	Red huckleberry	*Vaccinnium parvifolium* Smith	Shrub
	Salal	*Gaultheria shallon* Pursh	Shrub
	Salmonberry	*Rubus spectabilis* Pursh	Shrub

	Willow	*Salix* spp.	Shrub
	Red alder	*Alnus rubra* Bong.	Tree
	Western hemlock	*Tsuga heterophylla* (Raf.) Sarg.	Tree
	Western redcedar	*Thuja plicata* Donn.	Tree
Mosses	Big red stem	*Pleurozium schreberi*	Moss
	Broom moss	*Dicranum* spp.[1]	Moss
	~4-8 undetermined Peat mosses	*Sphagnum* spp.[1]	Peat moss
	Spread-leaved peat moss	*Sphagnum squarrosum*	Peat moss
Lichens	Antlered perfume	*Evernia prunastri*	Foliose
	Cetraria	*Cetraria orbata*	Foliose
	Hooded bone	*Hypogymnia physodes*	Foliose
	Lipstick cladonia	*Cladonia macilenta*	Foliose
	Parmotrema	*Parmotrema chinense*	Foliose
	Ragbag	*Platismatia glauca*	Foliose
	Tattered rag	*Platismatia herrei*	Foliose
	2 undetermined *Cladina*	*Cladina* spp.[2]	Fruticose
	2 undetermined *Usnea*	*Usnea* spp.[2]	Fruticose
	3 undetermined *Cladonia*	*Cladonia* spp.[2]	Fruticose

[1]*Requires microscopic analysis for positive ID to species.*

[2]*Requires chemical tests for positive ID to species.*

And Then...The beavers Arrived!

Figure 83- Beaver moving from Earth Sanctuary's Middle Pond to the Fen Pond. (c) Chuck Pettis

Earth Sanctuary is a prime habitat for beavers. When the previous owner, Robert Tank, bought the property in 1974, he reported, "There was a beaver lodge near the west end of the large fen pond, about 30 feet from the south shore. It was in good condition, but I never saw any activity around it and never saw any beaver in any of the ponds during the succeeding 26 years. The lodge gradually degraded over time."

Figure 84 - Western Red Cedars eaten by beavers! (c) Chuck Pettis

Fast forward to 2012 when beavers re-inhabited Earth Sanctuary. When the beavers arrived, they started eating many newly planted Western Red Cedars. I would walk by a tree that was 6" in diameter yesterday and today was a stump. So, we had to start fencing the Western red cedars by the ponds. We started with chicken wire fencing, but it immediately became apparent that that wasn't high enough as beavers can stand over three feet tall. The fences are now 3.5 to 4 feet tall.

Beaver Dams Became a Problem

Beavers became obvious in all three Earth Sanctuary ponds from the damming, tree felling, lodges, and beaver trails throughout. The two existing drainage devices at Earth Sanctuary known as weirs were implemented in the mid-20th Century. These devices consist of concrete-walled chambers in front of culvert pipes built to prevent overflow of the banks separating the middle and west ponds and between the west pond and the drainage ditch beside Newman Road. They were not built to prevent beaver damming.

Beaver damming requires a constant effort to keep the drains clear. Photo and video evidence of beavers in the act of damming was documented using motion-activated infrared trail camera technology. The beavers started plugging up the weirs, water control structures with culvert pipes between the ponds. Beavers are programmed to stop the sound of running water; that's why they make dams.

Do you know how difficult it is to unplug a culvert pipe that a beaver has skillfully plugged? Clearing a culvert pipe that has been stuffed by beavers is extremely difficult! (https://shorturl.at/bfA4f) Then we added a fence in front of the culvert pipe. Every day I would have to use a pitchfork to clear the fencing in front of the culvert pipe. Every night the beavers would return to "repair" their dam (https://shorturl.at/x2AVe).

So, we researched "beaver deceivers," ultimately ending up with state-of-the-art beaver deception technology, the Clemson Beaver deceiver[10]. The Clemson beaver deceiver consists of a white plumbing pipe with holes surrounded by a 3-foot-diameter fence to keep the beavers from plugging up the holes. The pipe extends out into the pond underwater and is held up by a float. The Clemson beaver deceivers eliminate the sound of running water and they work great!

To control water flow at Earth Sanctuary, mitigate beaver damming, and limit the need for frequent dam clearing, two Clemson beaver deceivers were constructed and installed to fit directly into the existing drainage system at Earth Sanctuary.

During late August 2013 the first Clemson beaver deceivers was installed between the middle pond and the west pond. A 10-foot section of 6-inch diameter septic pipe was used for the intake pipe and three 10-foot sections of 4-inch diameter septic pipe were used for the outflow pipe assembly. Two 90-degree angle clean-out joints were utilized in the outflow section with a short section of pipe between the joints to allow for the entire intake device to be submerged into the water and easily cleaned.

Figure 85 - Pond leveler before installation. (c) Earth Sanctuary.

A wooden dam with a 4-inch hole was cut out of plywood and secured into the existing weir using grooved 2x6s positioned on the sides of the weir. The last two feet of the outflow pipe assembly were inserted through the dam.

Figure 86 - Design of wooden dam with outflow pipe flow. (c) Earth Sanctuary

A large metal buoy was affixed to the end of the intake device out in the pond using a stainless steel chain to suspend the whole leveler in the water and avoid permanently mounting it to the bottom of the pond.

Figure 87 - West pond leveler after installation. (c) Earth Sanctuary

In late August 2014, a second pond leveler was installed between the west pond and Newman Road drainage.

Since the installation of the two Clemson beaver deceivers, there have been no problems with beaver plugging. Water levels in the middle pond and west pond have remained stable and no dam removal or maintenance has been necessary.

Where the Heart Finds Peace

Oh, Nature's sweetest balm, how fair thou art,
With birds that sing, and trees that reach the sky,
And ponds that shimmer in the sun's warm light,
And sacred spaces where the heart finds peace.

The birds, they twitter in the gentle breeze,
Their melodies are a symphony of joy.
A choir that lifts the spirit to new heights,
And fills the soul with peace and harmony.

The trees, they stand so tall and proud and strong,
Their branches reaching up towards the sky.
A haven for the birds that make their home,
A place of beauty that can never die.

And in the ponds, the yellow pond lilies bloom,
A world of life that dances in the sun.
A place of peace, a sacred, holy room,
Where all is one, and all is said and done.

Oh, Nature's sweetest balm, how fair thou art,
A haven for the soul, a work of art.

Questions for Group Discussions

These questions aim to provoke thoughtful discussion and deepen understanding of the principles, practices, and impacts of the Earth Sanctuary project.

- What do you think of the idea of a 500-year plan?
- What are the key steps in the "Design with Nature" ecological restoration methodology used at Earth Sanctuary?
- What strategies are used at Earth Sanctuary to enhance wildlife habitat? How do these strategies contribute to biodiversity and the health of the ecosystem?
- Have you had experience with beavers? What happened?
- What challenges have been encountered in the restoration and maintenance of Earth Sanctuary?
- Considering the long-term nature of the 500-year plan, what future challenges might Earth Sanctuary face?
- How can the principles and practices used at Earth Sanctuary be applied to other ecological restoration projects?
- How can communities and organizations adopt similar long-term ecological plans to benefit their local environments?

Chapter 11: **The Birds of Earth Sanctuary**

"While doing some quiet chanting yesterday in the Earth Sanctuary's spectacular Dolmen, a hummingbird came into the space and hovered for a long time like a blessing to my meditation. What an incredible sanctuary this is that even the birds feel safe and happy to join in the space.!"

"So many amazing moments, hard to pick just one. But looking into the eyes of a Barred Owl for 20 minutes was very special and moving. Thank you for making Earth Sanctuary!"

"We enjoyed watching a large covey of quails scurrying one at a time between their shrubby sanctuaries."

"I loved the sound of all the birds and seeing the content robins with bellies full of worms."

Birds of Earth Sanctuary, Freeland, Whidbey Island, WA (https://tinyurl.com/mup84ssj) - A video by Craig Johnson

Within the boundaries of Earth Sanctuary lies a rich and unique habitat that provides refuge for a diverse community of animals. The Earth Sanctuary's three ponds, and their associated wetland buffers occupy over 70% of Earth Sanctuary's land. They've been recognized for their importance as waterfowl habitats and designated as a "Habitat of Local Importance" by the Whidbey Audubon Society and the Island County Critical Areas Program.

Figure 88 - Female and male Wood Ducks. (c) Rick Collar.

On any given spring morning at Earth Sanctuary, you may see the local pair of nesting Osprey, adolescent and mature Bald Eagle, Barred Owl, Great Horned Owl, Wood Ducks, Hooded Mergansers, Tree and Violet-green swallows, and Great Blue Heron, among others.

Figure 89 - Osprey - what's for dinner? (c) Jann Ledbetter

The forested areas of the Earth Sanctuary provide habitat resources for more than 100 species of birds. The aquatic environment of the ponds provides a significant feeding ground for birds. The ponds' protected shorelines, the bog island, as well as the many tree snags (dead and dying trees) provide wonderful nesting locations for birds. Most of these species are perching birds (known for their singing), woodpeckers, hummingbirds, owls, ospreys, and hawks.

Figure 90 - Barred Owl on standing stone at the Cottonwood Stone Circle. (c) Chuck Pettis

Contributions to Conservation

Many thanks to Yvonne Palka and Frances Wood who in 2003 and 2004 conducted surveys for Earth Sanctuary identifying breeding birds using criteria established by the Audubon Society. And many thanks to Sarah Schmidt who reviewed this chapter and made a number of needed changes and edits, as well as updating the Bird Checklist (see end of Chapter).

See the 2003 breeding bird survey report

https://earthsanctuary.org/pdf/BreedingBirdSurvey2003.pdf

See the 2004 breeding bird survey report

(https://earthsanctuary.org/pdf/BreedingBirdSurvey2004.pdf)

Figure 91 - Hooded Merganser with chicks (c) Chuck Pettis

Life Around the Ponds

The ponds provide a wonderful habitat for a wide variety of bird species. Yvonne reports, "At the high point of baby ducks in mid to late June approximately 35 baby ducks were feeding and swimming on the central and west ponds. It looked and felt like a big playground for ducklings with attentive mamas swimming close by!"

Unfortunately, predation from eagles, owls, largemouth bass, bullfrogs, raccoons, and river otters often takes its toll on ducklings.

Figure 92 - Male Gadwall Duck. (c) P. Vishnevetsky

Yvonne continues "We focused a lot of our attention on the activity on the ponds themselves, since this is the unique part of Earth Sanctuary's habitat and the part that makes it a 'habitat of local importance.' The ponds, particularly the Middle Pond, is an extremely rich habitat for breeding ducks. It is quite secluded, and the surrounding forest is prime nesting territory for Wood Ducks and Hooded Mergansers. The vegetation on the shore and the islands is home to Mallards, Green-winged Teals, Cinnamon Teals, and Gadwalls. We saw many courting pairs of all of these species on the West and Middle ponds in early May."

Duck Species	Number of Broods	Dates Hatched	Approx. # Young Hatched
Mallard	2	May 5th, June 10th	14
Hooded Merganser*	**3-4	May 21st, June 27th	14
Wood Duck	3	May 21st, June 9th, June 27th	21
Green-winged Teal	1	June 28th	5
Cinnamon Teal	1	July 19th	4

Figure 93 - Table of species with dates and numbers of broods and young at hatching in 2003

Figure 94 - Cedar Waxwings on a branch on the Middle Pond. (c) Chuck Pettis

For breeding ducks, the ponds offer lots of good feeding as well as islands for the ground-nesting species. The many logs offer perches for the ducks to get out of the water and preen and warm up and still be safe from ground predators. This is very important for ducklings in particular.

Many other birds feed on the ponds. During May and June, large numbers of swallows (five species) and Cedar Waxwings are catching insects over the water. Kingfishers are successfully fishing, and Herons are feeding as well. Tree Swallows are nesting in the dead trees standing in the pond (they require nesting cavities). And undoubtedly, many bats are catching insects over the water at night.

Bird Nesting Boxes

Since early 2002, we have installed a multitude of bird nesting boxes, which are commonly known as birdhouses. Many thanks to the Whidbey Audubon Society for their monitoring and reporting work, and to the Boy Scouts and leaders of Troop 57 for cleaning, repairing, and filling nesting boxes with wood chips.

8 SHORELINES February 2023

Volunteer Corner:

Earth Sanctuary Nest Box Monitoring Project

We are so thrilled that Whidbey Island Boy Scout Troop 57 has volunteered to work on cleaning, repairing, and replacing the nest boxes used by cavity nesting ducks at the Earth Sanctuary.

Earth Sanctuary offers the perfect habitat, a wooded wetland, for cavity nesting ducks, especially, Hooded Mergansers, Bufflehead and Wood Ducks. Starting in early spring, late March, monitoring volunteers visit the Sanctuary weekly, walking the grounds and visiting the area where nest boxes have been established. The monitoring program runs from late March to early July, which encompasses the nesting and fledging periods for these cavity nesting ducks. Volunteers visit the three ponds in the sanctuary; West, Middle and Fen Ponds, and record all bird species they observe and the number of birds of that species.

The monitoring program goal is to record the species and numbers of cavity nest ducks observed, the number of chicks observed, and any eggs or egg-shells found around the area. A weekly tally of all bird species observed is collected and saved for a final report submitted to Whidbey Audubon Society by early August.

Please send an email to *contact.us@whidbeyaudubonsociety.org* if you are interested in getting involved in this monitoring project.

— Ann Casey and Jann Ledbetter

Photo from Earth Sanctuary website

Female Wood Duck

Dates to Remember

Please check the Whidbey Audubon Society Events Calendar and Listings on its website for more detailed information and how to enter Zoom for some of these upcoming events: *www.whidbeyaudubonsociety.org*. There also might be further

Figure 95 - Nest Box Monitoring Project. (c) Shorelines

Earth Sanctuary
Checklist of Birds – Updated August 2023 by Sarah Schmidt (106 species)

This daily field sheet lists the current status of species ever recorded from Earth Sanctuary. Species followed by "*" have been identified as breeding at Earth Sanctuary (using criteria established by the Audubon Society). Species followed by "**" have been identified as probably breeding at Earth Sanctuary. Please help keep this list current by reporting new species to Chuck Pettis at cpettis@earthsanctuary.org.

Waterfowl
- Cackling Goose
- Canada Goose*
- Trumpeter Swan
- Wood Duck*
- Cinnamon Teal*
- Northern Shoveler
- Gadwall*
- American Wigeon
- Mallard*
- Northern Pintail
- Green-winged Teal*
- Ring-necked Duck
- Bufflehead
- Common Goldeneye
- Barrow's Goldeneye
- Hooded Merganser*
- Common Merganser
- Red-breasted Merganser

Quail and Allies
- California Quail

Grebes
- Pied-billed Grebe**

Pigeons and Doves
- Rock Pigeon
- Band-tailed Pigeon*
- Eurasian Collared-Dove
- Mourning Dove**

Hummingbirds
- Anna's Hummingbird*
- Rufous Hummingbird*

Rails, Coots and Cranes
- Virginia Rail**
- Sora**
- American Coot

Herons and Allies
- Great Blue Heron

Vultures
- Turkey Vulture

Hawks and Eagles
- Osprey*
- Northern Harrier
- Sharp-shinned Hawk
- Cooper's Hawk
- Bald Eagle

Owls
- Great Horned Owl
- Barred Owl

Kingfishers
- Belted Kingfisher

Woodpeckers
- Red-breasted Sapsucker*
- Downy Woodpecker*
- Hairy Woodpecker*
- Pileated Woodpecker*
- Northern Flicker*

Flycatchers
- Olive-sided Flycatcher*
- Western Wood-Pewee*
- Willow Flycatcher*
- Hammond's Flycatcher
- Pacific-slope Flycatcher*

Vireos
- Hutton's Vireo*
- Warbling Vireo**

Corvids
- Steller's Jay*
- American Crow
- Common Raven

Chickadees
- Black-capped Chickadee*
- Chestnut-backed Chickadee*

- Martins and Swallows
  - Northern Rough-winged Swallow
  - Tree Swallow*
  - Violet-green Swallow*
  - Barn Swallow*
  - Cliff Swallow
- Bushtits
  - Bushtit*
- Kinglets
  - Golden-crowned Kinglet*
  - Ruby-crowned Kinglet
- Nuthatches
  - Red-breasted Nuthatch*
- Treecreepers
  - Brown Creeper*
- Wrens
  - House Wren
  - Pacific Wren* (formerly Winter Wren)
  - Marsh Wren*
  - Bewick's Wren*
- Starlings
  - European Starling*
- Thrushes
  - Townsend's Solitaire
  - Varied Thrush
  - Swainson's Thrush*
  - Hermit Thrush
  - American Robin*
- Waxwings
  - Cedar Waxwing*
- Old World Sparrows
  - House Sparrow
- Finches and Allies
  - Evening Grosbeak
  - House Finch*
  - Purple Finch**
  - Red Crossbill*
  - Pine Siskin*
  - American Goldfinch*
- New World Sparrows
  - Chipping Sparrow
  - Fox Sparrow
  - Dark-eyed Junco*
  - White-crowned Sparrow*
  - Golden-crowned Sparrow
  - Song Sparrow*
  - Lincoln's Sparrow
  - Spotted Towhee*
- Blackbirds
  - Red-winged Blackbird*
  - Brown-headed Cowbird*
  - Brewer's Blackbird
- Wood-Warblers
  - Orange-crowned Warbler*
  - Nashville Warbler
  - Common Yellowthroat*
  - Yellow Warbler*
  - Yellow-rumped Warbler*
  - Black-throated Gray Warbler
  - Townsend's Warbler
  - Wilson's Warbler*
- Cardinals and Grosbeaks
  - Western Tanager*
  - Black-headed Grosbeak*

A Family That Has Found Its Way

In the tranquil pond at Earth Sanctuary,
A pair of wood ducks reside,
Their feathers a vivid display of color,
As they gracefully glide.

In a nesting box on a nearby tree,
Their chicks just hatched with care,
As they climb out in feather puffs,
They fall to the ground, unaware.

But fear you not, they quickly recover,
And waddle their way to the pond,
To join their parents and all of the others,
In a community of which they are so fond.

The wood ducks are a sight to behold,
A family that has found its way,
In the peaceful waters of Earth Sanctuary,
Where nature's beauty is on full display.

Questions for Group Discussions

These questions should stimulate thoughtful discussion and deeper understanding of the chapter's themes and messages.

- Have you ever had a memorable encounter with wildlife? Share your experience and how it impacted you.
- Discuss the significance of Earth Sanctuary being designated as a "Habitat of Local Importance" by the Whidbey Audubon Society and the Island County Critical Areas Program.
- What roles do the nesting boxes play in supporting bird populations at Earth Sanctuary?
- How do community involvement and volunteer efforts contribute to the preservation and monitoring of bird habitats?
- How can individuals and communities contribute to similar conservation efforts in their local areas?

Chapter 12:
My Earth Sanctuary Investment

The ability to purchase Earth Sanctuary stemmed from my career in brand consulting. Over the course of my 50-year career, I was privileged to work in a wide variety of industries: high-tech, manufacturing, computer hardware and software, telecommunications, banking, restaurants, fashion, healthcare, Internet, retail, and service businesses. I also worked with numerous non-profit and spiritual/religious organizations.

One pivotal opportunity came in the 1990s when I consulted for several Internet companies, including The Outpost Network. This company was eventually acquired by InfoSpace, a high-flying Internet stock at the time. To complete the acquisition, The Outpost Network had to pay off its debt, offering shares for just a nickel each. I invested 10% of my net worth, purchasing $40,000 worth of shares, which eventually translated into 80,000 shares in InfoSpace.

A Spiritual Commitment

In 1999, while attending a month-long Tibetan Buddhist Lamdre retreat in Darjeeling, India, I visited an internet café and discovered that InfoSpace's stock price had skyrocketed. On the back of a retreat brochure, His Holiness Jigdal Dagchen Sakya had written a request for a retreat center in the Seattle area. Inspired, I approached him and pledged to create a retreat center.

The Birth of Earth Sanctuary

In 2000, I sold all my stocks. My wife, Claudia, encouraged me to do something meaningful with the money, and I purchased the four parcels that now comprise the 72-acre Earth Sanctuary.

Wayne Tippery, my broker, introduced me to the initial 57-acre property, which wasn't officially for sale. He advised, "Bring a machete," as the land was overrun with Himalayan blackberries and virtually impenetrable. The beauty and vibrancy of the three ponds captivated me, and I knew instantly that this was the perfect location. I later acquired three additional parcels:

1. A 10-acre plot that had been logged but not replanted.
2. The parcel and house at the corner of Newman and Emil roads, which became the Earth Sanctuary Retreat Center.
3. A small nearby parcel on a hill.

The total cost of these acquisitions was $817,514.21.

Early Vision: Eco-Resort

Initially, I envisioned an eco-resort and enlisted the expertise of Ludovic Laine, a Cornell University hospitality graduate, and Cheki Dev, Professor of Marketing at Cornell University. Two Cornell hospitality classes studied the feasibility of this idea. However, zoning regulations and limits on the number of rooms made the concept financially unviable. I pivoted to creating Earth Sanctuary as a sacred space sculpture garden, nature reserve, and retreat center—a model that has proven successful.

Sustainability and Self-Sufficiency

Today, Earth Sanctuary is self-sustaining. Income from visitors, the Retreat Center, and Memorial Trees covers expenses such as tree planting, landscaping, property taxes, new sacred spaces, advertising, insurance, and property management.

Future Plans for Earth Sanctuary

The Vision & Goals

- **Foster Biodiversity:** Create and protect a diverse and mature old-growth forest with maximum biodiversity, aligned with the Earth Sanctuary 500-Year Plan.
- **Spiritual Renewal:** Provide inspiring sacred spaces for personal renewal and spiritual growth.
- **Memorial Trees:** Ensure access to Memorial Trees for the foreseeable future.
- **Peaceful Environment:** Offer a sanctuary for mindfulness, prayer, and meditation.
- **Harmony Between Nature and People:** Prioritize the co-existence of nature and people.

- **Non-Denominational Retreat Center:** Maintain a spiritual retreat space that is inclusive of all traditions.

Strategies for the Future

- **Ecosystem-Based Design:** Use ecological and spiritual approaches to design and development, allowing the land to guide us in creating harmonious spaces.
- **Integration of Art and Spirit:** Combine restoration ecology with art, ceremony, prayer, and meditation.
- **Daily Spiritual Practices:** If possible, a meditator will walk the trails regularly, praying and performing spiritual practices.
- **Preserve Knowledge:** Document and preserve the plans, files, and history of Earth Sanctuary for future reference.

Land-Use Plan

- Dedicate the majority of Earth Sanctuary to open space, forests, wetlands, and natural habitats.
- Cultivate an old-growth forest and increase vegetation to maximize biodiversity.
- Establish a conservation easement upon my passing or when appropriate to ensure the land-use plan is enforced.
- Transition Earth Sanctuary management to a non-profit organization in partnership with a land trust.

Closing Reflections

Earth Sanctuary represents the culmination of my spiritual aspirations and professional achievements. By merging ecological restoration with spiritual intention, I hope to create a legacy that will inspire and benefit future generations, providing a space where the beauty of nature and the sacredness of life coexist in harmony.

Questions for Group Discussions

- What was Chuck Pettis's initial vision for Earth Sanctuary, and how did it evolve over time?
- How does Chuck's approach of creating a sacred space sculpture garden, nature reserve, and retreat center reflect his values and goals?

- Discuss the feasibility study conducted by Cornell University for the eco-resort idea. What challenges were identified, and how did these challenges shape the final vision for Earth Sanctuary?
- How does Chuck's decision to focus on spiritual and ecological goals over financial returns align with the overall mission of Earth Sanctuary?
- Reflect on the importance of establishing a conservation easement and a non-profit organization to manage Earth Sanctuary in the future.

Part 2

Earth Sanctuary's Sacred Spaces

Chapter 13:
The Earth Sanctuary Cottonwood Stone Circle

My Love for Stone Circles

I love stone circles. I built my first stone circle in 1977 in Ithaca, New York - the Ellis Hollow Stone Circle. Oriented to the summer solstice sunrise, the Ellis Hollow Stone Circle is still humming today with spiritual power.

Figure 96- Ellis Hollow Stone Circle Summer Solstice Sunrise. (c) Chuck Pettis

Figure 97 - The Ellis Hollow Stone Circle in 1997. (c) Lydia Pettis

I built my second stone circle in 1995 – the Mutiny Bay Stone Circle. Its stones face west and orient to the Summer Solstice, the Spring and Fall Equinoxes, and the minor and major declinations of the moon. It was blessed and consecrated by H.H. Jigdal Dagchen Sakya and H.E. Dagmo Kusho Sakya. Prayers were done for the beings, e.g., insects, harmed or killed during its construction.

The Vision Behind the Cottonwood Stone Circe

The Cottonwood Stone Circle at Earth Sanctuary was designed to be a place for prayer, meditation, and spiritual discussion. It is a place of power, protection, healing, and safety.

Figure 98- Cottonwood Stone Circle. (c) Michael Stadler.

Built-in 2003, the Cottonwood Stone Circle is 40 feet in diameter and is located in an open meadow between the Middle Pond and the Fen Pond. It is surrounded by a ring of Black Cottonwood trees. The Cottonwood Stone Circle has an outer circle of 12 standing stones 11' high and is surrounded by 1,400+ native plants. Using special geometry, the circumference of the stone circle is exactly 120 feet. This dimension is also built into the geometry and metrology of the Mutiny Bay Stone Circle.

Figure 99- Welcome to the Cottonwood Stone Circle! (c) Chuck Pettis

In 2006, many stones were added to the original circle of twelve stones. I love tall stones, so three very tall Columbia River basalt columns were added within the stone circle, along with several sitting stones and a ceremonial fire pit. These inner stones are arranged artistically to complement the symmetry and formal geometry of the twelve outer stones.

Videos by Bre Pettis showing the installation of the center stones:

Part 1 - Building the New Cottonwood Stone Circle (https://tinyurl.com/ydj73mam)

Part 2 - Building the New Cottonwood Stone Circle (https://tinyurl.com/2b4tze97)

Part 3 - Building the New Cottonwood Stone Circle (https://tinyurl.com/5n7wt3eb)

Part 4 - Building the New Cottonwood Stone Circle (https://tinyurl.com/bded589p)

Figure 100- Lydia Pettis in the Callanish Stone Circle. (c) Chuck Pettis

Inspired by the Callanish Stone Circle in Scotland, a line of standing stones was added on the west and the east axes, and two outlier stones were placed to dramatize the north-south orientation of the stone circle.

The Spiritual Energy of the Cottonwood Stone Circle

At the center of the Cottonwood Stone Circle is a ley line power center that holds the strong emotions and experiences of those who have visited here. Many people have commented on the feelings they get in this sacred space. One woman once told me that the Cottonwood Stone Circle is one of two places on Earth where she feels safe. A group of children told me they had just had the best conversation with their father that they had ever had. Several groups of breast cancer survivors have visited the Cottonwood Stone Circle. They stand by each of the twelve standing stones and remember their emotional experience in each of the past twelve months. A woman grieving the loss of her mother touched her head against one of the stones and immediately felt comforted by the ancient presence emanating from it.

Figure 101- H.E. Dagmo Kusho Sakya, monks and group. (c) Chuck Pettis

Ceremonies and Blessings

Two Tibetan Buddhist fire ceremonies were held here, led by His Holiness Jigdal Dagchen Sakya, now passed. A Shinto priest gave a blessing ceremony and sprinkled sake on all the standing stones. Father Bill Johnson, a Jesuit priest, celebrated a version of Teilhard de Chardin's "Mass of the World" in the Cottonwood Stone Circle. And Klaw-osht, a Native American Shaman who created the two Medicine Wheels at Earth Sanctuary held a ceremony here. The place is filled with spiritual power and blessings.

Short Instructional Meditation Video for the Cottonwood Stone Circle

(https://tinyurl.com/e4ux9s25)

Experiencing the Cottonwood Stone Circle

To experience the power and energy of the Cottonwood Stone Circle, stand in the center. Close your eyes and feel your sensations. What do you feel? Then, do a prayer or meditation of your choosing. Or try this meditation.

Close your eyes and watch your thoughts. A thought is anything you name or identify, such as a bird sound, the feeling of wind on your cheek, or memories of where you've been or what you are going to do later. Our brain is filled with synapses all connected. One thought leads to another. In meditation, the goal is is create spaces of calmness between the thoughts. To practice, start by counting your thoughts. When a thought arises, count it, 1, 2, 3, and so forth. See how many thoughts you have in 60 seconds. Just by watching your thoughts over time, you will start to have fewer thoughts. The fewer thoughts you have, the happier you will feel.

Gratitude

Thank you for your meditations, prayers, and blessings at the Earth Sanctuary Cottonwood Stone Circle.

Figure 102- Tall Standing Stones in the Cottonwood Stone Circle rising to the heavens. (c) Michael Stadler

Figure 103- Ring of Brodgar - Orkney Islands.
One of the inspirations for the Cottonwood Stone Circle. (c) Chuck Pettis

Symbolic Sentinels

In lands of ancient lore and mystery,
Where time's embrace keeps secrets safe and sound,
There stand in stoic grace, with history,
The stone circles, their stories yet unbound.

Beneath the stars, a timeless ballet plays,
Each stone a dancer in this sacred rite,
In patterns formed, as nights turn into days,
They weave the tales of worlds beyond our sight.

Symbolic sentinels, these stones arise,
In circles, arches, mighty portals tall,
They reach towards the heavens, seeking skies,
Connecting Earth to cosmos, large and small.

In cycles spun, like seasons' endless flow,
The stones remind us of life's ebb and tide,
The birth, the growth, the waning's gentle woe,
In nature's rhythm, journeys we abide.

The solstice sun, with purpose, finds its mark,
Aligning shadows cast by stone and sphere,
An ancient calendar, a guide in the dark,
Guiding spirits, pilgrims, year by year.

The unity of life, the bond we share,
These circles speak of kinship, love's embrace,
A reminder that all life is rare,
In every land and culture, every race.

Through ages past, these circles stood their ground,
Witnessed generations rise and fall,
Their mystery endures, profound, unbound,
Enigmas etched in stone, a timeless thrall.

So, as we ponder their symbolic ways,
Let stone circles guide us through the night,
To grasp the essence of these ancient days,
And find in unity, our common light.

Chapter 14:
The Earth Sanctuary Dolmen

What is a Dolmen?

Physically, a dolmen is a megalithic monument constructed of upright stones supporting a large horizontal slab.

Figure 104 – A Neolithic dolmen at Carnac, France was the inspiration for the Earth Sanctuary dolmen. (c) Chuck Pettis

The Earth Sanctuary dolmen is the first dolmen to be built in the U.S. during modern times. It serves as a place for meditation and reflection where visitors can center themselves and discover an intense, unmistakable connection with the earth.

Figure 105 - The Earth Sanctuary dolmen (c) Michael Stadler

Spiritual Significance of the Dolmen

Spiritually, Earth Sanctuary's dolmen is intended to serve as a meditation room, within which, surrounded by the strong vibrations of the lichen-covered Montana sandstone, you can center yourself and discover an intense, unmistakable connection with the earth. The dolmen represents, in the Hindu expression, the "cave of the heart," a deep place of interior knowing that lies at the center of each one of us.

Like a womb, the cool, sheltering space of the dolmen provides a nourishing and protected place for reflection. It leads us back to the feminine space of the earth.

Constructing the Dolmen

Figure 106- The dolmen construction team (c) Chuck Pettis

The stones (hard Montana sandstone) of the dolmen came from Marenakos Rock Center. I asked each stone if they wanted to be at Earth Sanctuary. Then I made clay models of the stones to determine the best way to fit them together onsite. The dolmen was installed on a weekend by Albert Gabelein and his son Steven, as well as my sons, Bre and Morgan, and myself. Fun!

A Ley Line Power Center

The dolmen marks the first ley line power center at Earth Sanctuary. A ley line power center combines the yin calming effect of underground water with energetic yang ley lines. The resulting consciousness field holds the strong emotions, prayers, and meditations that you and people like you experience here. In this way, you leave blessings for those who come here in the future. Meditating and praying here is a meritorious spiritual practice.

Blessings and Consecrations

Figure 107 - H.H. Jigdal Dagchen Sakya conducted the Tibetan Buddhist fire ceremony, assisted by H.E. Dagmo Kusho Sakya, Dhungsey Mati Sakya, and Lama Nyima Lama. (c) Chuck Pettis

The dolmen is a blessed and consecrated environmental artwork. In 2002, His Holiness Jigdal Dagchen Sakya conducted a Tibetan Buddhist fire ceremony to bless and consecrate this sacred space. Since then, other high lamas, native leaders, and grandmothers have also blessed this power center.

Figure 108 - Vi Hilbert at the Dolmen Photographer unknown

For example, Vi Hilbert, a revered Upper Skagit/Salish Grandmother, teacher, and keeper of Native American wisdom, once visited the dolmen. She offered words of gratitude to the spirits of the megalithic stones of

Earth Sanctuary's dolmen. Thanking them for their strength and serenity, she added "You will be here long after we have been gone. You will be here to do work for generation after generation."

How to Meditate in the Dolmen

The dolmen is a perfect place to do a 60-second calm-abiding meditation focusing on your breath. Here's how to do it.

1. Go inside the dolmen.
2. Standing or sitting, close your eyes and focus on your breath.
3. Count your breaths. An inhale followed by an exhale is one breath.
4. How many breaths do you have in 60 seconds?

Short Instructional Meditation Video for the Earth Sanctuary Dolmen

The Dalai Lama's Insight on Breath Meditation

My son, Bre, and I attended His Holiness the Dalai Lama's 80th birthday in New York City. The Dalai Lama spoke for 25 minutes on the benefits and importance of counting your breaths. He encouraged people to try and count to 21 breaths without getting distracted. He said when you can do that, see how high you can count!

The Landscaped Approach to the Dolmen

Figure 109 – Landscaped approach to the Dolmen. (c) Michael Stadler

The approach to the Dolmen has been carefully landscaped by Dan Borroff and features a series of viewing platforms. This table-shaped stone structure and the surrounding stones are built of 20 tons of Autumn Leaf hard sandstone from Montana. The stones range in height from five to seven feet. Underneath each of the standing stones are buried colored marble and quartz crystals, as a spiritual crystal complement to the Autumn Leaf sandstone

Gratitude

Thank you for your meditations, offerings, prayers, and blessings at the Earth Sanctuary dolmen.

A Portal to the Veil

In realms where whispers of antiquity roam,
Where legends dance upon the ancient loom,
There lie in stately grace, amid the gloam,
The dolmens, guardians of forgotten tomb.

A mystic tale, their weathered stones confess,
Of souls departed to the great unknown,
Where spirits soar in realms of boundless guess,
In dolmens' arms, their final journey was sown.

Beneath the arch of stars that twinkle bright,
These mighty sentinels of ages past,
They cradle secrets veiled in ancient light,
In granite arms, the stories still hold fast.

Each dolmen stands, a portal to the veil,
Where past and present waltz in cosmic rhyme,
Their silent presence speaks of timeless tale,
Where memory and time entwine sublime.

The sun's embrace, a dance of shadows play,
Through shifting seasons, ever they endure,
Their solemn grace, a tribute to the day,
An ode to lives that once adorned this moor.

In hallowed ground, they rise with strength untold,
Defying years, unwavering in stance,
As history unfolds, their tales unfold,
In quiet dignity, they watch and trance.

Let us, with reverent hearts, draw near to glean,
The wisdom etched in dolmens' sacred face,
And in their whispers from the past convene,
To touch the echoes of a bygone grace.

For in these ancient stones, a treasure gleams,
A connection to our roots, a sacred span,
Embracing all the wonders of our dreams,
The dolmens stand immortal in the plan.

Chapter 15:
The Earth Sanctuary Labyrinth

A Journey of Reflection and Renewal

"When we got to the labyrinth, we noticed that the stone path was covered with windfall and alder leaves. We were delighted to find a broom nearby, which enabled us to clean up the path so we could hone the space and contribute our energy and intention to it."

"The walk through the labyrinth walked me back through time to let the past go. The journey out gave me hope and direction for the future."

The Purpose and Power of Labyrinths

Labyrinths are used worldwide as a way to quiet the mind, calm anxieties, recover balance in life, enhance creativity, and encourage meditation, insight, self-reflection, and stress reduction.

Labyrinths Through History

Labyrinths appear in a variety of forms and patterns throughout history. Unlike mazes, which incorporate tricks and dead ends and are designed to confuse you, labyrinths have only one path leading to the center and back out again. Labyrinths are designed to calm and center you.

Symbolism

The labyrinth is a universal symbol, appearing in various forms across cultures:

- **Christian Tradition:** A symbol of pilgrimage.
- **Celtic Tradition:** A magical tool for communication with the other world.
- **Hopi Tradition:** Representing the sacredness of nature, reflecting the spiral of plant tendrils.

Constructing the Earth Sanctuary Labyrinth

Figure 110- Earth Sanctuary labyrinth just after placing the Pennsylvania Bluestone pavers. Dan Borroff (designer) and Dave Schmidt (landscaper) in the center. (c) Chuck Pettis

The labyrinth at Earth Sanctuary is located on a tree-sheltered small peninsula that extends into the Middle Pond. This peninsula was once covered by 16-foot-high Himalayan blackberries that took a lot of work to remove! As you walk the trail to the labyrinth look around. We planted all the plants and trees here except for the very large and older trees.

The Earth Sanctuary labyrinth was constructed as a tool for walking meditation, with the idea that walking into the labyrinth's center and out again can be experienced as a metaphor for life's journey or as a way to access your higher self. Its design is based on the three-circuit labyrinth, the most ancient labyrinth form, dating to around 2500 B.C.

- **Diameter:** 32 feet.
- **Path Materials:** Pennsylvania Bluestone flagstones, dry set on a compacted sand base.

Evolution of the Salal Hedge

From 2002 to 2023, a salal hedge bordered the Pennsylvania Bluestone pathway. In 2023, the salal was removed because it was diseased and not functioning as an effective hedge. After discussion, we have decided to leave the labyrinth as it is without a hedge.

Figure 111- Earth Sanctuary Labyrinth just after planting Salal hedge. Photo (c) Victoria Scarlett

Figure 112 - Salal hedge after a rain shower. (c) Chuck Pettis

Figure 113 - Earth Sanctuary labyrinth just after the Salal had been removed by Dave Schmidt's Edmond Collage class. (c) Chuck Pettis

The Spiritual Elements of the Labyrinth

One of the keys to creating a place of power is to juxtapose opposites. The opposites cancel out, creating sacred spaces that evoke transcendental experiences. Like some other sacred spaces at Earth Sanctuary, the labyrinth is laid out with a North-South alignment. Buried under this labyrinth are grapefruit-sized Herkimer diamonds and black tourmaline that function as a spiritual complement to the Pennsylvania bluestone. Herkimer diamonds have metaphysical properties including healing and increased spiritual energy. Black tourmaline clears and protects people from negative thoughts and anxieties.

Walking the Labyrinth

There is no right or wrong way to walk a labyrinth. Here are a few suggestions:

- **Be Present:** Let go of thoughts and focus on your experience.
- **Seek Insight:** Walk to the center while holding a question or intention.
- **Reflect:** Use sacred texts, scripture, or mantras.

- **Offer Blessings:** Send love to someone in need at the labyrinth's center.

At the center of the Earth Sanctuary labyrinth is a ley line power center. When you reach the center, close your eyes, and feel your sensations. Fill your heart with love and send that love to someone who needs extra love as a blessing.

Short Instructional Meditation Video for the Earth Sanctuary Labyrinth (https://tinyurl.com/e4ux9s25)

The Native American Prayer Stone

On the edge of the Middle Pond near the Labyrinth entrance is a Native American prayer stone. Visitors can pray or meditate here following the instructions, including removing shoes and offering tobacco.

~~~

The Spiritual Qualities of a Labyrinth

In mazes intricate, a labyrinth resides,
A sacred realm where souls may journey deep.
Its winding paths, a tapestry that guides,
To solace and peace, where spirits find keep.

With every step, a pilgrimage unfolds,
A dance of introspection and release.
A chance to shed the burdens we hold,
And find the inner stillness, inner peace.

In labyrinthine turns, we lose our way,
Yet find ourselves amidst the winding road.
The twists and turns become a gentle sway,
That leads us to the truths we must behold.

A sacred symbol of the human quest,
The labyrinth invites us to invest
Our hearts and minds in paths that intertwine,
To seek the sacred, in our souls, align.

Within its patterned lines, we find a grace,
A chance to glimpse the depths of our own being.
As ego fades, and spirit takes its place,
We're touched by something greater, more foreseeing.

The center beckons like a guiding light,
A beacon of divinity's embrace.
~~~

We step with faith into the sacred night,
Embracing mystery, with every pace.

For in this sacred space, we are restored,
Our spirits are lifted; souls are set alight.
The labyrinth, a gateway to the Lord,
Unveiling truths that bloom in depths of night.

So let us walk this labyrinth of grace,
And find the sacred in each winding trace.

~~~

Journey Deep Inside

In winding paths of ancient mystic lore,
A labyrinth stands, a sacred realm's door.
Its intricate design, a soulful guide,
Inviting seekers to journey deep inside.

Step by step, upon its hallowed ground,
The seeker's spirit starts to be unbound.
A dance of faith, a meditative trance,
In labyrinth's embrace, the soul may enhance.

With every turn, a whisper from the past,
Resonating echoes, secrets amassed.
The winding walls, a mirror to the heart,
Reflecting truths, where inner worlds impart.

In labyrinth's center, a sacred space,
Where time and space unite in an embrace.
A moment's stillness, a presence divine,
The seeker finds their soul and purpose align.

Oh, labyrinth, a labyrinth of the soul,
A pathway to the realms that make us whole.
Guide us, inspire us, in your gentle grace,
Unveiling mysteries as we embrace.
~~~

Chapter 16: **The Earth Sanctuary Medicine Wheel**

"I found the Native American medicine wheel to be deeply moving. To know that we've acknowledged that this land was stolen from those of the past, and that Earth Sanctuary is working with the tribes to renew the relationship to the land with healing intentions felt powerful and rich. This set strong intentionality for the restoration of the land, both ecologically and spiritually."

"This was my first time to pray at the Medicine Wheel. I appreciated the clear instructions and could feel the sacredness of the spaces. When I finished my ceremony, I felt renewed and cleansed. It was very powerful and I'm so grateful for all the healing opportunities and spaces at Earth Sanctuary."

"The sacred sites yielded profound meditative experiences, especially the Prayer Stone and the Medicine Wheels…it is lovely how you've posted protocols for use to make them accessible."

Introduction to the Medicine Wheel

The Medicine Wheel, one of the most powerful and cherished symbols of Native American belief, is an ancient place of prayer and contemplation sacred to Native Americans, symbolizing the totality of existence. At Earth Sanctuary, the Medicine Wheel is a sacred space designed to amplify the power of prayers and connect you with all the Grandmothers, Grandfathers, Mother Earth, Moon, Sun, Sky, and the Creator, in the eternal circle of life.

Origins of the Earth Sanctuary Medicine Wheel

The Medicine Wheel was created by Klaw-osht, a shaman of the Nuu-chan-nulth tribe of Vancouver Island. Klaw-osht appeared at Earth Sanctuary from the very beginning. I would see him walking by the Dolmen, not knowing who he was. When I got to know him, he told me stories about the

Gray Whales and how he would go out with the men from his tribe. They would call the whales, and one would offer itself up to them. He showed me an aerial photo of his home and the inlet looked like a whale.

By agreement with Klaw-osht, no photos of the Medicine Wheel itself are allowed. However, **Figure 114** shows the four colors at the center of the Medicine Wheel.

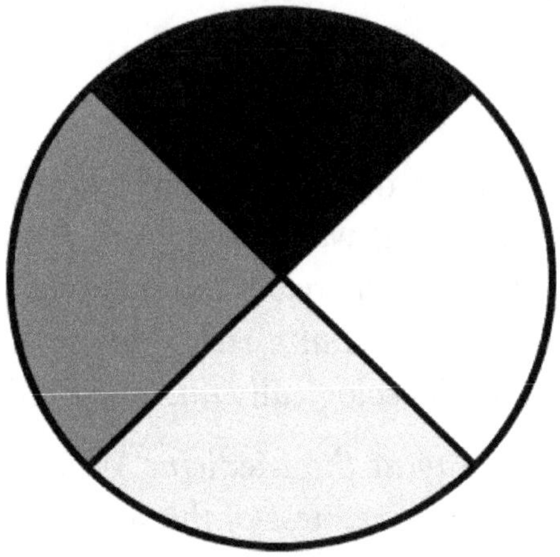

Figure 114 - Image showing the four colors in the center of the Medicine Wheel.

Design of the Medicine Wheel

The Medicine Wheel is built on 20 tons of white sand. The wheel itself has four colors for each of the four directions. There are prayer flags holding tobacco around the Medicine Wheel that are renewed annually with a Native American blessing and consecration ceremony.

Location of the Earth Sanctuary Medicine Wheel

The Medicine Wheel is located near the Cottonwood Stone Circle, accessible via the Wetland Trail. To reach it, walk up the steep hill carefully. From the top, the Medicine Wheel overlooks the Cottonwood Stone Circle and the Fen. If you want to catch a glimpse of the Medicine Wheel without entering, walk past the bench on the left-hand side, where you can see the center wheel with its colors of yellow, red, black, and white.

This sacred space is lovingly cared for by a volunteer. Due to the legal agreement with Klaw-osht, we ask for respect in adhering to the no-photography rule.

Experiencing the Medicine Wheel - Native American Prayer Protocol

To fully experience the Medicine Wheel, Native American prayer protocol should be observed. Directions, protocol, and materials for using the Earth Sanctuary Medicine Wheel are provided both at the site. These instructions are printed on a laminated sheet for visitors to use within the Medicine Wheel.

Summary of the Process

In brief, here's what you will do:

1. **Smudging for Purification:** Begin by smudging yourself for purification. During high fire danger in the summer, we ask that you visualize the smudging process instead.
2. **Walking the Wheel:** Holding a small amount of tobacco in your left hand, walk around the Medicine Wheel. At each of the four directions, say your prayers.
3. **Offering Tobacco:** Leave your tobacco on the spherical Grandmother Stone as an offering before departing the Medicine Wheel.

Donations and Gratitude

Donations of tobacco and smudge are greatly appreciated and bring blessings to you. Thank you for your prayers and blessings at the Earth Sanctuary Medicine Wheel.

Closing Thoughts

The Medicine Wheel is a sacred space that embodies connection, respect, and reverence. By following the prayer protocol and honoring the traditions it represents, you become part of the eternal circle of life that the Medicine Wheel symbolizes.

~~~

Here is a poem written by Klaw-osht for the Medicine Wheel.

We share our tears with you.

Oh Great Spirit who dwells in our hearts and in the Universe,

Lead us through the path of sorrow.

Give the family peace and understanding of Life's journeys end.
~~~

Be merciful and have pity on our hearts.
Our lives are so short here, walking upon Mother Earth's surface.
Open our hearts and eyes to all the blessings you have given;
The time she shared with each and everyone.
Take this woman who is Sister, Daughter, Auntie, Wife, Mother and
Grandmother, with open arms, into the land of her ancestors,
the beautiful land of happiness and loving music.
May the healing music of Grandfather Sky and Grandmother Earth
forever echo in our hearts.
When night falls, may Grandmother Moon shine her cool, soft, silver
Blanket over her place of rest and peace.
Oh Great One of all our relations,
Please hear our prayer.

Klaw-osht

Nuu-chan-nulth

Douglas G. David

The Native American Spiritual Leader Who Created the Earth Sanctuary Medicine Wheels

Chapter 17:
The Earth Sanctuary Baby Gray Whale Medicine Wheel

"During my two-day stay, I enjoyed many serene walks, deep meditations, and powerful stone formations. My favorite was a Gray Whale Medicine Wheel, which is the skeleton of a baby whale. After quieting by the awe of this, as requested, I started singing AND dancing joyfully!"

Figure 115- The Earth Sanctuary Baby Gray Whale Medicine Wheel. (c) Chuck Pettis

The Baby Gray Whale Medicine Wheel is an authentic Native American medicine wheel created and blessed by Klaw-osht, a shaman of the Nuu-chan-nulth tribe from Vancouver Island. This sacred site serves as a shrine for all water beings, with a special connection to whales.

Stories of the Gray Whales

Klaw-osht shared stories of his life on Vancouver Island and the traditions of his people. He recounted how, as a young boy, he would accompany his father, grandfather, and other men of the tribe to the ocean. Together, they would meet with the gray whales, forming a deep spiritual connection with these majestic beings.

My wife and I live on the west side of Whidbey Island on Puget Sound. One day a gray whale appeared in front of the beach. Filled with wonder, I called Klaw-osht to tell him about it. Tragically, the next day the whale beached and died. Klaw-osht visited the site to offer prayers and honor the whale's spirit.

About a month after that, Klaw-osht called me and told me to come to his house. Beside his driveway was a baby gray whale skull. Curious, I asked Klaw-osht where he got it. He replied, "I prayed for it."

This profound gift found its home at Earth Sanctuary. A small group of us took the baby gray whale skull up the hill to its current location. It's heavy! Those of us who participated in this journey will never forget the powerful ceremony that Klaw-osht performed to bless the Baby Gray Whale Medicine Wheel. As part of this sacred ritual, Klaw-osht also buried an eagle at the site, enhancing its spiritual significance.

Visiting the Medicine Wheel

When you visit the Baby Gray Whale Medicine Wheel, please approach with gentleness, quietude, and respect. Klaw-osht offers a unique request: think of your favorite song and sing it silently to yourself while standing next to the medicine wheel. According to him, the whales will hear your song when they pass by Whidbey Island.

A Place of Connection

The Baby Gray Whale Medicine Wheel is more than a shrine; it is a beacon of connection between humans, nature, and the sacred world of the whales. Your songs, meditations, prayers, and blessings contribute to this profound spiritual space.

Thank you for honoring this sacred place and joining in its purpose at Earth Sanctuary.

Chapter 18:
The Earth Sanctuary Fen Stone Circle

"We found the energy in the Fen Stone Circle to be very powerful. We are in appreciation of the dedication that has been gifted to this place. It is very evident that the earth here is holding the loving energy for all that visit and for all mankind."

"I didn't know what to expect. This place is overwhelming, all the beauty, the solitude. I had quite an experience in the Fen stone circle, one I will never forget. I can't wait to come back with more family and friends so they can also experience the magic of this sacred place. Thank you."

Figure 116 - The Earth Sanctuary Fen Stone Circle. (c) Chuck Pettis

Overview of the Fen Stone Circle

The Fen Stone Circle, overlooking the serene Fen Pond, measures 16 feet in diameter and is made of eight standing stones, each reaching up to seven feet in height. These thin Llenroc stones weigh 600-1,000 pounds each and come from the Finger Lakes Stone Quarry in Ithaca, New York. Intriguingly, the name Llenroc is Cornell (l-l-e-n-r-o-c) spelled backward, symbolizing a unique connection to its place of origin.

A Connection to the Past

The Fen Stone Circle is deeply connected to my first stone circle, the Ellis Hollow Stone Circle, built in 1977 and later expanded in 1997 with larger stones, all made of Llenroc. Wanting to create a link between Earth Sanctuary and the Ellis Hollow stone circle, I arranged for Llenroc stones to be shipped from Ithaca to Earth Sanctuary. When the stones arrived on a pallet, some were broken. We repaired these stones by pinning and gluing them back together, a process that mirrors moments in life when I had to piece myself back together after making mistakes. Each time, I emerged stronger and wiser, just as these stones now stand firm in their repaired state.

Alignments and Energetic Intentions

The Fen Stone Circle's design includes four pairs of stones aligned with cardinal and solstice points: true North-South, the Summer Solstice sunrise and sunset, and the Winter Solstice sunset. Adding to its spiritual depth, a second, hidden stone circle made of Rose quartz, sourced from the Black Hills of South Dakota, lies buried within the Llenroc circle. This inner circle was created to generate an energetic field of welcome and safety.

A Transformative Experience

The Fen Stone Circle has proven to be a site of profound energy and wonder. One memorable instance involved a retreatant from the Retreat Center who called me with a remarkable story. He asked, "What are you doing with this stone circle?" He recounted how, while standing within the circle, he removed his glasses and discovered he could see perfectly. However, once he left the stone circle, he needed his glasses again. This experience exemplifies the circle's potential for miraculous, transformative moments.

A Space for Blessings and Reflection

The Fen Stone Circle has also been a site for sacred ceremonies.

Figure 117 - H.E. Dagmo Kusho and others circumambulating the Fen Stone Circle. (c) Chuck Pettis

Figure 118 - H.H. Jigdal Dagchen Sakya blessing the Fen Stone Circle. (c) Chuck Pettis

To fully experience the circle's power, stand at its center. Close your eyes, tune into your sensations, and take a moment to feel its energy. Engage in a prayer or meditation of your choosing or try the following practice.

A Meditation for Love and Healing

Begin by bringing up the feeling of love within you. For me, this feeling arises instantly when I think of my grandchildren. Let love fill your heart.

Now, think of someone who could use extra love and send your feelings of love to them as a blessing and spiritual gift.

Thank you for your meditations, prayers, and blessings at the Earth Sanctuary Fen Stone Circle. Every moment of reflection and intention enriches the energy of this sacred space.

Chapter 19:
The Earth Sanctuary Ley Line Sculpture

"It's truly remarkable how the ley lines make such a difference in the energy vibrating in certain areas."

"This visit I especially liked the ley lines experience! I am thankful for the opportunity to recharge in a dedicated spiritual space with all the wonderful art/spirit places envisioned and created by Chuck Pettis."

The most potent ancient monuments and sacred spaces around the world have one thing in common: the presence of special earth energies, which have the power to alter and uplift human consciousness. Earth Sanctuary follows this tradition. Its sacred spaces are located over ley-line power centers and are networked with ley lines.

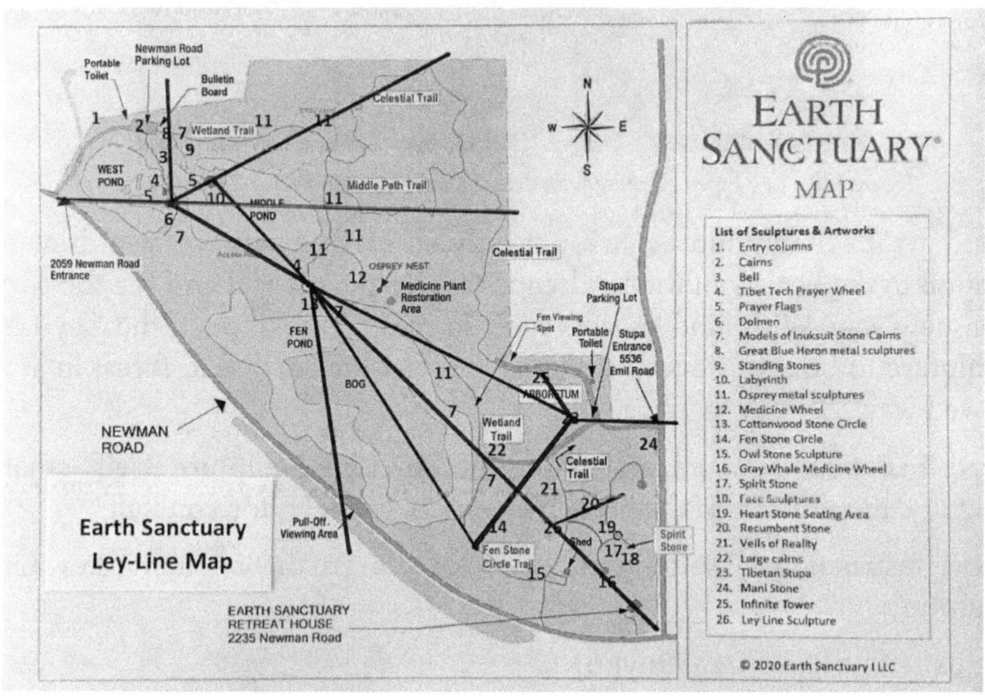

Figure 119 - The Earth Sanctuary Ley Line Map (c) Chuck Pettis

Driftwood and Small Earth Mound

The burnt driftwood poles located beside a rounded earth mound found their way to Earth Sanctuary. The driftwood pieces were selected one by one to form the grouping that marks the significant ley power line running from the Earth Sanctuary Retreat Center to the Cottonwood Stone Circle.

Connecting Art and Landscape

Figure 120 - Ley Line Sculpture with deer in the background – (c) Barbara Philipp

Land art inevitably must mark not only an earth event but also draw attention to the beauty of the landscape to help people feel and see the earth in a new way. With land preservation in mind, I positioned the Ley Line Sculpture to be experienced viscerally, to feel the earth and forest in a renewed way.

Earth Sanctuary's sacred spaces and the Ley Line Sculpture itself actually magnify and illuminate the human experience. Feel it for yourself!

Short Instructional Meditation Video for the Earth Sanctuary Ley Line Sculpture

(https://tinyurl.com/za99nyk6)

Figure 121 – The Ley-Line Sculpture. (c) Chuck Pettis

Experiencing the Ley Line Sculpture

To experience the ley line, stand between the driftwood poles, close your eyes, and feel the sensation of a ley line. Then, sit or stand in the middle of the diamond-shaped space at the end of the line: the power center. Again, close your eyes and feel the sensation of a ley line power center. Allow yourself to feel inspired and uplifted by the earth's energy.

Chapter 20:
The Earth Sanctuary Compressed Cube Tensegrity Sculpture

Inspiration from Buckminster Fuller

When I attended Carnegie Mellon University in the late sixties, I became inspired by Buckminster Fuller's vision of making "Spaceship Earth" work for 100 percent of humanity. I also fell in love with his geodesic domes. Building domes, zomes, and inflatables for the 1970 Earth Day was a transformational experience. I was entranced by how these structures have special acoustic and psychological properties.

Inventor R. Buckminster Fuller holds up a □Tensegrity□ sphere in Philadelphia April 9, 1979, which he describes as a bundle of rods and cables that can be made into a giant sphere so light and strong it can support a floating space station a mile in diameter. The rods don□t touch each other but are held in place by tension on the cables. The integrity of the sphere is such that repairs of any section does not disturb the rest of the geodesic structure. (AP Photo/Bill Ingrahim)

Figure 122 - Buckminster Fuller inspired me in many ways. What a visionary! Photo licensed from AP.

Learning Tensegrity

After graduating from Carnegie Mellon, I went to Buckminster Fuller's design school at Southern Illinois University. One of the professors there was Anthony Pugh who taught tensegrity and wrote the now out-of-print excellent book, "an introduction to tensegrity. (https://a.co/d/aro1s3X)"

What is Tensegrity?

Tensegrity (https://tinyurl.com/2s4hcbe6) refers to a structural system formed when a set of discontinuous compression components interact with a set of continuous tension components to define a stable and balanced volume in space.

Figure 123 - "Free Ride Home" tensegrity sculpture by Kenneth Snelson at Storm King Art Center. (c) Chuck Pettis

A Long-Standing Dream: Creating the Compressed Cube Tensegrity Sculpture

The Compressed Cube Tensegrity Sculpture represents the culmination of my early studies with Buckminster Fuller (https://tinyurl.com/8fp6rxht) and my admiration of Kenneth Snelson's (http://kennethsnelson.net/)

tensegrity artworks. I have always dreamed of creating my own tensegrity sculpture and this vision came to life.

Figure 124 - Compressed Cube Tensegrity Sculpture. (c) Michael Stadler

In early 2021 I started making a few tensegrity models, an icosahedron, and then progressing to a cube. When I compressed the cube, it transformed into a unique and captivating shape that to my knowledge no one had created before.

Figure 125 - None of the pipes touch. (c) Michael Stadler

The Floating Puzzle

The Compressed Cube tensegrity sculpture consists of four 12' long by 5" in diameter aluminum tubes connected with 1/8" cables. These cables are arranged in such a way that the metal tubes do not touch each other, creating the illusion that the structure is floating in space. Despite this appearance, the structure is locked into a stable and balanced form.

This sculpture embodies a moment of stasis and electrifying. It is a puzzle of form floating apart and at the same time in perfect balance. Race-car red powder coating was the icing to further electrify the tensegrity forces.

Chapter 21: The Earth Sanctuary Stupa

"Holy objects like the Earth Sanctuary Stupa have the power to bring immeasurable spiritual good and blessings."

"Earth Sanctuary is an ideal location to transform your mind to a brighter and more compassionate state. When I engaged in the morning meditation, I would take time to circumambulate around the blessed stupa to say my mantras. I discovered the local wildlife will subtly participate in your experience. As a woodpecker hammered away on the side of a nearby tree, I imagined it was the playful sound of a Tibetan damaru drumming in support."

"Circumambulating the Stupa was meaningful and healing, calming, and centering. Thank you for creating and maintaining this sacred space."

"The stupa was a rejoicing moment of stability and heavenly broadcast of endless bliss."

The stupa is the most important Buddhist monument and is a sacred space, representing Buddha's holy body, speech, and mind, as well as the path that one follows to become a Buddha and enlightened being.

Figure 126 - The Earth Sanctuary Stupa (c) Michael Stadler

Figure 127 - The Usnisa Vijaya statue. Usnisa Vijaya is one of the three long-life Buddhas. (c) Chuck Pettis

The Role of the Stupa in Spiritual Practice

According to Buddhist teachings, a stupa helps calm the mind, radiates spiritual blessings, amplifies prayers, promotes long life, attracts wealth, helps cure illness and disease, and plants the seeds of enlightenment.

We invite everyone who visits to walk around the stupa clockwise with reverence, prayer, or meditation; this is called circumambulating the stupa. This is more than a ritual; it is a profound spiritual practice.

The Transformative Power of Circumambulation

If we have difficulties in life and want to achieve success, we need to purify the negative karma that caused these difficulties and create merit for future success and happiness. Walking around the stupa clockwise while saying prayers, mantras, or thinking spiritual thoughts can help find solutions to many problems.

Suggested Practices for Circumambulating the Stupa

Here are some suggested prayers and chants to use when circumambulating the stupa:

- **Pray:** "May all beings be happy. May all beings be free of suffering."

- **Personal Prayer:** Say a prayer from your own spiritual or religious tradition. The stupa will amplify your prayers.
- **Visualization:** Imagine white light emanating from the stupa, totally illuminating and purifying you from sickness and negative karma.
- **Chant:** Recite the most popular Tibetan Buddhist mantra: *Om Mani Padme Hung*. This is the mantra of Chenrezi, the Bodhisattva of loving-kindness and compassion.

Figure 128 - H.H. Jigdal Dagchen Sakya and monks who conducted the blessing and consecration ceremonies. (c) Chuck Pettis

Resources for Visitors

If you visit the Stupa at Earth Sanctuary, you will find a handout on the bulletin board with a complete explanation of the stupa and its significance. This resource is designed to deepen your understanding and enhance your spiritual experience.

Thank you for your meditations, prayers, and blessings at the Earth Sanctuary Stupa.

Chapter 22:
The Earth Sanctuary Veils of Reality Sculpture

Figure 129- Veils of Reality Sculpture (c) Chuck Pettis

The Veils of Reality Sculpture consists of two four-foot squares that juxtapose opposites, a black steel square plate on one side, and white moiré fabric on the other. The use of paired opposites creates a balancing force that facilitates transcendental experience – moments beyond objective explanation become comprehensible. This principle of balancing opposites is integrated into all the sacred spaces at Earth Sanctuary.

The Moiré Effect: A Gateway to Quantum Mechanics

The moiré effect is a visual perception that occurs when a set of lines or dots is superimposed on another set, creating fascinating and intricate geometric patterns. These patterns occur naturally, arising from the interference between two overlaid designs. Interestingly, moiré patterns have significant implications for quantum science applications, especially at the nanometer scale. In a profound way, observing the moiré patterns on this sculpture offers a glimpse into quantum mechanics in action, bridging art, nature, and science.

A Sacred Ceremony

As part of a sacred Tibetan Buddhist fire ceremony conducted at the Earth Sanctuary stupa, His Holiness Jigdal Dagchen Sakya, head lama of Sakya Monastery of Tibetan Buddhism, led Tibetan monks in the creation of a sand mandala on the sculpture's steel plate. This intricate mandala was consecrated through a sacred Vajrapani ceremonial fire was lit on top. The steel plate, now imbued with the blessings of Tibetan Buddhism, has been transformed into a new art form—a union of spiritual tradition and modern minimalist design.

Figure 130- The Vajrapani file ceremony. (c) Chuck Pettis

Who is Vajrapani? Vajrapani protects the practitioner from harm and removes the obstacles in their path with the thunderbolt held in his right hand. He is depicted in a wrathful form, and he is alternatively known as the Lord of Secrets and the Demon Subduer.

Minimalism and Self-Reflection in Art

Minimalist art strives to take what is less and make it more. The Veils of Reality Sculpture invites viewers to engage deeply with its simplicity, encouraging them to see their lives in a new light. By contemplating the black steel plate, viewers are encouraged to discover fresh insights and perspectives within themselves.

Capturing the Unseen

Figure 131 - What will you see when you look into the black steel plate? (c) Chuck Pettis

In minimalist art, the artist takes what is less and makes it more. By looking intently at a pure piece of art, we can see ourselves in a new light. 'Veils of Reality' encourages viewers to view their lives with new insight and perspective.

Figure 132 - If you visit the Veils of Reality sculpture, have someone take a photo of you. The camera picks up images that the eye doesn't see. (c) Chuck Pettis

Visitors are encouraged to stand before the black steel plate and observe what they see. Each person's experience is unique, as the sculpture reflects something different for everyone. Interestingly, photographs of individuals standing before the plate often capture subtle elements that are invisible to the naked eye. This phenomenon adds another layer of wonder to the experience.

Ultimately, the Veils of Reality Sculpture invites us to focus, reflect, and view ourselves and the world in a transformative new way. Take your time, look deeply, and enjoy the journey.

Chapter 23:
The Earth Sanctuary Infinite Tower

Connecting Heaven and Earth

Figure 133- The Infinite Tower artwork of spiraling geometric shapes connects heaven and earth at Earth Sanctuary. (c) Chuck Pettis

Figure 134 - The Infinite Tower. (c) Michael Stadler

Standing at 24 feet tall, the Infinite Tower is a column of interlocking equilateral triangles that creates a deceptively fluid visual effect. The design symbolizes a profound connection between Earth and the sky. The choice of white, a color representing clarity and purity, allows the sculpture to reflect shadows and shapes dynamically, enhancing its visual allure. The nearby trees and snags also resonate with this connection, joining the earth and sky in harmonious unity. Furthermore, the numerology within the design is deeply intentional: the number eight (eight triangles in each octahedron) signifies balance and cosmic order, while the number seven (seven stacked octahedrons) represents spirituality and harmony.

Figure 135- View of Infinite Tower from Below (c) Chuck Pettis

A Tribute to Buckminster Fuller

The Infinite Tower is my homage to Buckminster Fuller, one of the greatest visionaries of the 20th century. My admiration for Fuller began during my time at Carnegie Mellon University in Pittsburgh, where I had the privilege of hearing him speak multiple times. His works, especially the *World Design Science Decade* documents, profoundly influenced my thinking by demonstrating how the world's resources could be utilized for the benefit of all humanity.

Inspired by Fuller's vision, I built numerous models of polyhedra—shapes with many faces—as well as geodesic domes, zomes, and inflatables. Decades later, in 2016, I began experimenting with stacking polyhedra to create something unique. After exploring various forms, I concluded that octahedrons offered the ideal structure. Using a MakerBot Replicator 2 Desktop 3D Printer, I 3D-printed a set of octahedrons and stacked eight of them to create a model. When I presented the concept to my artist wife, Claudia, her affirmation of its beauty solidified my decision to turn the design into a sculpture.

Crafting and Installing the Tower

Collaborating with Matthew Swett of Taproot Architects, I developed a detailed blueprint for the 24-foot column of interlocking equilateral triangles. Metal artist Tim Leonard of Heavy Metal Works brought the design to life, expertly constructing the sculpture.

During the installation of the Infinite Tower, I placed a quartz crystal beneath it. The moment the tower was secured, I felt the presence of a ley line power center, activated by the Landscape Angel. This alignment of spiritual and physical elements elevated the Infinite Tower beyond its physical form, making it a beacon of energy and harmony.

Figure 136 - A cool way to take a photo at the Earth Sanctuary Infinite Tower. (c) Chuck Pettis

Capturing the Magic of the Infinite Tower

Visiting the Infinite Tower offers a unique opportunity to experience its magic firsthand. For an unforgettable photo, try these creative angles:

1. Take a selfie with the top of the tower framed dramatically behind you.
2. Lie on the concrete pad and have a friend photograph you looking up, with the tower soaring above.

These perspectives capture not just the artistry of the Infinite Tower but also its profound connection to the heavens and the earth.

Chapter 24:
The Earth Sanctuary Peace Garden

Figure 137 - Caitlin Sullivan created the Earth Sanctuary Peace Garden (c) Chuck Pettis

Located next to the Earth Sanctuary Dolmen, Caitlin Sullivan, an Ambassador Girl Scout, created the Earth Sanctuary Peace Garden. For her Girl Scout Gold Award Project, she created an accessible place where people with mobility issues can feel peaceful.

Caitlin Sullivan's Vision

Caitlin says, "I was inspired to create the Peace Garden because many people in my community have mobility issues, including my grandparents and friends from work. I really wanted them to have a place to feel peaceful, and a place in nature to get away from the stress of daily life. Located beside the Dolmen, this is also the only area with seating and a view of the West Pond, one of Earth Sanctuary's three scenic ponds."

The Girl Scout Gold Award Project

Figure 138 - Caitlin Sullivan receiving her Girl Scout Gold Award. (c) Celia Sullivan

"The Girl Scout Gold Award is all about making a lasting impact on your community," Caitlin explained. "I live in a community that has many retired people in it, and many of them have trouble moving around. Earth Sanctuary is a place I enjoy going to to feel peaceful and to take a break from the craziness of the world. The people in my community, many of whom are my friends, would have trouble enjoying the peacefulness of Earth Sanctuary because there aren't many places to sit and enjoy nature. Since Earth Sanctuary is one of my favorite places, I wanted to find a way to expand its reach to more people in my community."

Community Support and Creation

Caitlin's project was supported by generous donations from Hanson's, a local lumber yard and hardware store, and Venture Out Nursery. Her Girl Scout troop and family organized a work party to bring her vision to life. "Even as we were creating the space, people would come through with encouraging stories and excitement about my project," Caitlin recalled.

A Lasting Impact

"Now that it is done, and I'm away at college, I continue to hear about the good it is bringing to Earth Sanctuary's visitors from Chuck," Caitlin added. Her work has left a lasting legacy, ensuring that visitors with mobility challenges can enjoy the beauty and tranquility of Earth Sanctuary's West Pond.

Chapter 25:
The Cairns of Earth Sanctuary
A Symbolic Journey Through Time and Spirit

The Origins and Significance of Cairns

Throughout Earth Sanctuary there are numerous small cairns, or mounds of stones. The word "cairn" originates from the Scottish Gaelic "càrn," meaning a heap of stones. In the remote corners of the world, amidst the rugged landscapes and natural wonders, you may come across a peculiar sight - a stack of stones, carefully balanced upon one another. These artful formations, known as cairns, have been a part of human history for millennia. Though seemingly simple, cairns hold deep spiritual qualities that resonate with individuals across cultures and time.

Figure 139- Cairn near Dolmen (c) Chuck Pettis

Practical Beginnings and Significance of Cairns

Initially, these stone stacks served practical purposes such as wayfinding and marking trails in vast wildernesses, helping travelers navigate safely through unfamiliar terrain. Over time, the significance of cairns evolved beyond their utilitarian role to encompass profound spiritual meanings.

Figure 140- Cairn at Fen Viewpoint points to prayer flags in the distant trees. (c) Chuck Pettis

Connection to the Earth

Cairns have an inherent connection to the Earth. As we gather stones, we are reminded of our primal bond with nature. Stones ground us in the present moment. The act of placing one stone upon another is an intimate dialogue with the land, fostering a sense of harmony and communion. Cairns symbolize our reverence for the Earth, honoring the sacredness of our interconnected existence.

Cultural Interpretations and Symbolism

In various cultures, cairns have served as navigational aids, guiding travelers through treacherous terrain, or marking significant points along pilgrimage routes. For example, the stone monuments of the Inukshuk (pronounced in-ook-shook) of northern Canada, Alaska, or even Greenland have multiple meanings. These cairns are vital for navigating the barren landscape and may also indicate a food source, serve as a warning marker, or mark a sacred area. The Inukshuk have developed the most comprehensive cairn "vocabulary." At Earth Sanctuary, models of some Inukshuk cairns have been created to align with Earth Sanctuary cosmology.

Figure 141- Cairn at trail junction (c) Chuck Pettis

Cairns as Life's Metaphor

Symbolically, cairns represent our journey through life—a metaphorical path filled with transitions, challenges, and personal growth. Cairns stand as beacons, offering reassurance and guidance as we navigate our inner landscapes just as they guide us on remote hiking trails such as the trail from Snow Lake to Lake Vivian in the Enchantment Lakes of Washington.

The Meditative Act of Building Cairns

Building cairns is a contemplative process that demands mindfulness and presence. Engaging in this activity can become a form of meditation, allowing individuals to clear their minds of distractions and connect with the present moment. The act of selecting stones, balancing them carefully, and witnessing the gradual creation of the cairn engenders a state of centeredness.

The very act of stacking stones to create a cairn requires patience, focus, and a sense of balance. The action mirrors our quest for equilibrium within ourselves. As we carefully select and arrange each stone, we become attuned to the delicate dance of stability and harmony. Cairns remind us to seek balance in our thoughts, emotions, and actions, encouraging us to find alignment between the physical, mental, and spiritual aspects of our being.

Cairns in Ritual and Spiritual Practice

Cairns have also been employed in spiritual rituals and ceremonies as sacred markers, signaling the presence of something profound or divine. They are used to mark the boundaries of sacred spaces, such as shrines or ritual sites, and function as conduits between the mundane and the transcendent.

As you enter the Newman Road parking lot at Earth Sanctuary, you will see a variety of cairns surrounding the parking area. By serving as vessels of spiritual energy, cairns become focal points for meditation, prayer, and connection to the divine.

Ancient Roots and Global Practices

In ancient India cairns were the forerunners of stupas, Buddhist symbols of enlightenment. In the Himalayas today and in many other places around the world, cairns are constructed as markers along shrine routes, footpaths, and places for offerings. Travelers in Tibet find Mani Stones almost everywhere, in monasteries, beside villages, crossing, along paths, and on mountains. Lama Migmar, Sakya Monastery's resident artist, painted the universal mantra, Om Mani Padme Hum, on the stone by the Stupa entrance in traditional colors and style.

Figure 142 - Mani Stone by Lama Migmar (c) Chuck Pettis

Reflections on Cairns

Cairns are more than mere stacks of stones; they are powerful symbols that transcend cultural boundaries. They remind us of our connection to the Earth, our journey through life's transitions, and the importance of balance and alignment. Cairns invite us to reflect on the spiritual essence within ourselves and the world around us, beckoning us to embrace the profound wisdom hidden within their humble forms.

Conclusion: The Enduring Presence of Cairns

Cairns, with their simple yet profound presence, have found their way into the spiritual fabric of human experience. Beyond being mere stone stacks, they carry deeper meanings that touch the souls of those who encounter them. As we build and encounter cairns, we tap into the inherent spirituality of these ancient structures, forging a link between the natural world and the human spirit. Cairns stand as a testament to our capacity for connection, guidance, and reflection, reminding us that in the vast wilderness of life, we are never truly alone.

Chapter 26:
The Earth Sanctuary Retreat Center

Testimonials from Earth Sanctuary Retreatants

"Earth Sanctuary is truly a sacred place. The retreat space feels like home. A home for the compassionate heart. Sitting in this space connects me to my heart and the hearts of so many before me. I came here to do heart-centered work. And I leave today feeling the support of land and a loving spirit. Thank you and many blessings!"

"The trails and sacred spaces provided a peaceful setting to reflect, meditate and appreciate nature and all her beings. I also enjoyed experiencing other sacred traditions. What an amazing place! I am so thankful it exists for all beings to experience as well as for preserving open space and creating habitat for all creatures. I look forward to coming again!"

"I visited the sacred sites – especially the Cottonwood Stone Circle and the Native American Medicine Wheel – daily. I found these places very comforting, and I felt encircled by revitalizing energy. I felt a quieting and calming from the first step onto the Wetland trail, and this has deepened consistently during my time here. This has been a truly healing and renewing experience for me. I am so glad I came! Every step here was on a sacred site."

"Earth Sanctuary is the perfect retreat from city noise and crowds. I hope to be back soon!"

"I have released my pain, my burdens, my sufferings. I carry this experience as a strength. I have been filled with love, patience, and gratitude. This is what I seeked when I came here. This is now my Sacred Space."

"Thanks for your retreat center/trail/art/place-based spiritual land. Loved it. Great trail system! Really enjoyed the pond and wetlands. The waterfowl are captivating and calming. Appreciate your work of intention and care for the land. Hope to return someday with more time."

"Thank you for sharing this quiet, peaceful retreat house and the miraculous land! I've had time away to meditate; pray, and have those prayers enhanced by the prayer wheels; commune with mother earth, the tall trees, and many beautiful animals. My heart feels renewed so I can take light back to the world."

"A great experience. I feel full. This was the time-out I needed to re-center my inner self and being. I'll be back again."

"This was just the perfect place for my personal retreat. I was able to hit a reset button and reconnect to nature while I was here."

"Wonderful space for our retreat. Thank you for sharing your sacred space grounds with us! So beautiful and calm. Namaste. Blessed."

"Much needed rest and rejuvenation! Getting back to what is most important – what is real. Thank you so much for this peaceful home."

"I came with an open heart and left with a full heart. I didn't find any particular answer but found the questions I wanted to ask. Thank you."

"I'm so glad I came here for four days. Every step I took on the path was to a sacred site, from the first step to the ones I will be taking as I continue my journey."

"We set up a silent meditation retreat here, and it was the perfect environment. The house provided regular meditation sits; the grounds provided spaces for walking reflection and the sanctuary overall is ideal for contemplative time. Energetically, it's restorative and healing."

Plan a Spiritual Retreat for Yourself

Consider a spiritual getaway at the Earth Sanctuary Retreat Center, a single-family house available for individual and small-group retreats.

Figure 143 - Exterior photo of the Earth Sanctuary Retreat Center. (c) Michael Stadler

Accommodations and Features

The Earth Sanctuary Retreat Center can accommodate 1-6 people overnight for personal renewal, relaxation, retreat, meditation, and peaceful reflection. A self-contained facility, those renting it have exclusive use of the entire Retreat Center.

The spacious living and meditation room provides stunning views of the surrounding trees and yard, creating a tranquil atmosphere for inner reflection.

Figure 144 - The spacious living and meditation room overlooks surrounding trees and yard. (c) Michael Stadler

Fully Equipped Kitchen

The kitchen is well-stocked with pots, pans, and utensils, allowing guests to prepare meals comfortably during their stay.

Figure 145- The kitchen is well supplied with pots, pans, and utensils. (c) Michael Stadler

Comfortable Bedrooms

Each of the two bedrooms is equipped with a single bed. Guests will also find a curated selection of books on meditation and spirituality to enhance their retreat experience.

Figure 146 - Each of the two bedrooms contains a single bed. Books on meditation and spirituality are available at the Retreat Center. (c) Michael Stadler

Location and Surroundings

The Retreat Center is located at the southeast corner of the Earth Sanctuary property, adjacent to the 72-acre nature reserve and sculpture garden. Guests have access to:

- Two miles of trails
- A diverse array of birds and wildlife
- Land art and sculptures
- Sacred spaces, including stone circles, labyrinths, a stupa, medicine wheels, and a dolmen

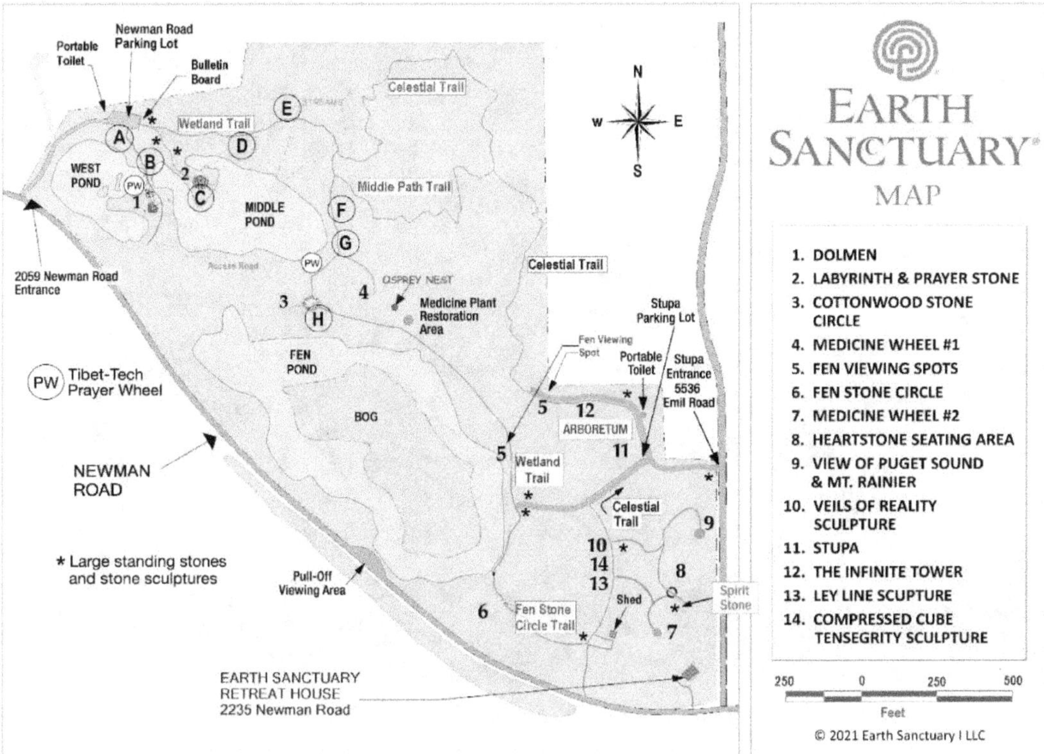

Figure 147 - The Retreat Center is located at the Southeast corner of Earth Sanctuary. (c) Earth Sanctuary I LLC.

Planning Your Visit

To learn more, visit the Earth Sanctuary website (www.earthsantuary.org) and click on Retreat Center. The site includes an availability calendar, photos of the Retreat Center, and details about accommodations, items to bring, directions, and more. When you are ready to make a reservation, fill out and submit the Online Reservation Form. Celia Sullivan, Earth Sanctuary's Retreat Center Caretaker, will confirm your reservation and then Chuck Pettis will send you a request for payment via PayPal.

Figure 148 - The Retreat Center is also a place for group retreats. (c) Chuck Pettis

A Legacy of Spirituality

The Earth Sanctuary Retreat Center has been hosting spiritual retreatants for over 20 years. Over 1,000 people have experienced a spiritual retreat at the retreat center. The collective spirituality of its retreatants permeates not just the Retreat Center, but the entire Earth Sanctuary property, making it a truly blessed and sacred place to do your spiritual work and spiritual practices.

Chapter 27:
Planning Your Visit to Earth Sanctuary

Welcome to Earth Sanctuary

You are invited to come and explore one of the most lovingly created eco-art sites in the world!

Earth Sanctuary combines exemplary earth stewardship with art and spirit to create a sanctuary for birds and wildlife, plus a peaceful and magical place for personal renewal and spiritual connection.

Figure 149- A lovely oasis. The beautiful Fen Pond at Earth Sanctuary. (c) Michael Stadler

What to Experience at Earth Sanctuary

Reclaim your spirit in Earth Sanctuary's beautiful forests and reflective ponds.

Explore:

- **Stone Circles**: Sacred spaces imbued with ancient wisdom.

- **Sculptures and Sacred Spaces**: Walk the labyrinth and amplify your prayers at the Native American medicine wheel.
- **Buddhist Stupa**: Learn about an authentic and spiritually significant structure.
- **Nature Reserve Trails**: Enjoy the abundance of birds while walking the two miles of serene trails.

Visiting Hours and Entry Fees

Earth Sanctuary is open to visitors every day of the year during daylight hours. The fee for visiting is **$7 per person**, which can be paid in the following ways:

- **Cash or Check**: Place payment in one of the provided envelopes and deposit it in the metal box near the bulletin board at one of the entrances.
- **Digital Payment**: Use Venmo or PayPal by scanning the QR codes on the bulletin boards.
- **Prepayment Online**: You can pay in advance by clicking "Day Pass" on the EarthSanctuary.org website.

Important Note

- **No Dogs Allowed**: While we love dogs, they can disturb Earth Sanctuary's fragile environment and wildlife. Please leave them in your car if you bring them.

Parking Information

Newman Road Parking Area (2059 Newman Road, Langley WA 98260)

- Park facing the hill in this designated area.
- If this lot is full, drive to the Emil Road entrance.
- Remember to pay at the bulletin board first before proceeding.

Emil Road Parking Area (5536 Emil Road, Langley, WA 98260)

- To get here from Newman Road, go to the end of the driveway, turn left, and in 0.6 miles, make another left onto Emil Road.
- The entrance, marked with prayer flags, has parking options near the bulletin board, as well as additional parking down or up the hill.

- Remember to pay at the bulletin board first before proceeding.

Guided Tour Guides and Maps

Printed tour guides with maps are available in literature holders at each bulletin board to help you navigate the nature trails and all the sacred spaces and sculptures. Explore:

Sacred Space Tour (https://tinyurl.com/4h2rnxxw),

Sculptures Tour (https://tinyurl.com/mdfkf238),

Naturalist Tour (https://tinyurl.com/23twzj2z))

If you have limited time, consider this two-step visit:

1. Start at the Newman Road parking lot to visit the Dolmen and Labyrinth.
2. Drive to the Emil Road entrance to visit the sculptures, Fen Stone Circle, and Cottonwood Stone Circle.

Closing Invitation

Please visit Earth Sanctuary soon - I look forward to meeting you on the trails and in one of the sacred spaces!

--Chuck Pettis

In Closing

Earth Sanctuary has become a deeply spiritual environment where people experience life-changing, transformational, magical moments. The concept of "Sacred Ecology," which combines relationships between all forms of life and spiritual beings, is the guiding philosophy behind Earth Sanctuary's ongoing efforts to create peace, happiness, and ecological balance. May this new book, *Secrets of Sacred Ecology*, help you create a nurturing, loving, and intimate relationship with nature.

Resources

Books

Spiritual Ecology: The Cry of the Earth edited by Llewellyn Vaughan-Lee. Second edition, The Golden Sufi Center, 2016.

Earth & Soul: Reconnecting Amid Climate Chaos by Leah Rampy. Bold Story Press, 2024.

Sacred Nature: Restoring Our Ancient Bond with the Natural World by Karen Armstrong. Anchor Books, 2023.

The Practice of the Wild – Essays by Gary Snyder. Counterpoint, 1990.

Ecology and Religion by John Grim and Mary Evelyn Tucker, Island Press, 2014

Sacred Ecology – Fourth Edition by Fikret Berkes. Routledge, 2018.

Spiritual Ecology: Reading the book of nature and reconnecting with the world by Rudolf Steiner. Rudolf Steiner Press, 2008.

Summoned by the Earth by Cynthia Jurs. Prospecta Press, 2024.

Becoming Supernatural: How Common People Are Doing the Uncommon by Dr. Joe Dispenza. Hay House, 2017.

Zen and the Art of Saving the Planet by Thich Nhat Hanh. HarperOne, 2021.

Secrets of Sacred Space: Discover and Create Places of Power by Chuck Pettis. Llewellyn Publications, 1999.

The Pattern of the Past by Guy Underwood. Pitman Publishing (January 1, 1971)

The Old Straight Track by Alfred Watkins. Abacus, 1988

View Over Atlantis by John Michell. Thames and Hudson, 1983

The Kingdom of the Gods by Geoffrey Hodson. The Theosophical Publishing House, 1976.

To Hear the Angels Sing: An Odyssey of Co-Creation with the Devic Kingdoms by Dorothy Maclean, Lindisfarne Press, 1994

Call of the Trees by Dorothy Maclean, The Lorian Association, 2006

The Book of Wilding: A Practical Guide to Rewilding, Big and Small by Isabella Tree, Bloomsbury Publishing, 2023

The Mind of Plants - Narratives of Vegetal Intelligence. Edited by John C. Ryan, Patricia Vieira, and Monica Gagliano. Synergetic Press, 2021

The Hidden Life of Trees; What They Feel, How They Communicate by Peter Whohlleben. Greystone Books, 2016.

Life After Life by Raymond Moody. HarperOne, 2015

On Life After Death by Elizabeth Kubler-Ross. Celestial Arts, 2004

Mind Beyond Death by Dzogchen Ponlop. Snow Lion, 2008

Peaceful Death, Joyful Rebirth by Tulku Thondup. Shambhala, 2006

The Three Levels of Spiritual Perception: A Commentary on The Three Visions

by Deshung Rinpoche. Wisdom Publications, 2003.

Crystalline Communion, Mineral Properties for Healing and Integration by Collen Marquist and Jack Marquist. Earthlight, 1987.

Wheel of Great Compassion: The Practice of the Prayer Wheel in Tibetan Buddhism by Lorne Ladner. Wisdom Publications, 2000.

The Silent Language by Edward T. Hall, Anchor, 1973

The Hidden Dimension by Edward T. Hall, Anchor, 1990

Shared Vision: The Second American Bauhaus by Al Gowan. Merrimack Media, 2012

Websites

Earth Sanctuary https://earthsanctuary.org/

Videos

Short Instructional Meditation Video for the Earth Sanctuary Dolmen.

Short Instructional Meditation Video for the Earth Sanctuary Labyrinth.

Short Instructional Meditation Video for the Cottonwood Stone Circle.

Short Instructional Meditation Video for the Earth Sanctuary Ley Line Sculpture.

Intelligent Trees

Clearing a culvert pipe that has been stuffed with mud by beavers:

Every night the beavers would return to "repair" their dam

Discover Earth Sanctuary - a video by Whidbey Telecom

Other Resources

Prayer wheels and prayer flags may be purchased from:

- Dharmashop: https://www.dharmashop.com/
- Etsy: https://www.etsy.com/
- Amazon: https://www.amazon.com/

Here is a link to a presentation on The Trees of Earth Sanctuary

See the 2003 breeding bird survey report

See the 2004 breeding bird survey report

Endnotes

1 At Earth Sanctuary, when we had a trash can, people filled it up. I took the trash can away and after that, no one left trash.

2 Excerpted from a Sakya Monastery class on Calm Abiding Meditation that I have taught for many years.

3 *Zen and the Art of Saving the Planet* by Thich Nhat Hanh. HarperCollins, 2012. Page 62-66.

4 Original illustration by Carrie Westfall (c)1999 Llewellyn Publications.

5 Original illustration by Carrie Westfall (c)1999 Llewellyn Publications.

6 https://www.shambhala.com/snowlion_articles/om-mani-padme-hum-dalai-lama/

7 Om symbolizes the pure exalted body, speech, and mind of a Buddha. Mani, meaning jewel, symbolizes the altruistic intention to become enlightened for the sake of sentient beings. Hum symbolizes the indivisible unity of method and wisdom.

8 Wheel of Great Compassion: The Practice of the Prayer Wheel in Tibetan Buddhism by Lorne Ladner, Wisdom Publications, 2000, 42.

9 *Call of the Trees,* page 16.

10 The Clemson Beaver Pond Leveler by Gene W. Wood and Larry A. Woodward, Clemson University. https://seafwa.org/sites/default/files/journal-articles/WOOD-179-187.pdf

www.ingramcontent.com/pod-product-compliance
Lightning Source LLC
LaVergne TN
LVHW060033210926
848481LV00013B/3850